Standards-Based Math

Grades 3-4

By
Jillayne Prince Wallaker

Published by Instructional Fair • TS Denison
an imprint of
McGraw Hill Children's Publishing

Author: Jillayne Prince Wallaker
Editors: Sara Bierling, Mary Rose Hassinger, Stephanie Garcia
Cover Illustration: Jeff Von Kanegan

Children's Publishing

Published by Instructional Fair • TS Denison
An imprint of McGraw-Hill Children's Publishing

Send all inquiries to:
McGraw-Hill Children's Publishing
3195 Wilson Drive NW
Grand Rapids, Michigan 49544

Standards-Based Math—grades 3–4
ISBN: 0-7424-0215-0

3 4 5 6 7 8 9 PHXBK 08 07 06 05 04 03

The McGraw-Hill Companies

Introduction

Standards-Based Math is based on the most recent standards from the National Council of Teachers of Mathematics. The ten proposed NCTM standards, which are a description of what an existing curriculum should enable students to know and do, can be divided into two categories. The Content standards are those five that state the skills areas that students should learn. These are Number and Operations, Algebra, Geometry, Measurement, and Data Analysis and Probability. The Process standards, which guide instructors in planning lessons that enable students to acquire knowledge and achieve success, are the second set of standards. These standards are Problem Solving, Reasoning and Proof, Communication, Connections, and Representation.

For the purpose of this book, the Content standards have been highlighted, and each page is labeled with its appropriate standard and skill. The Process standards are woven into the exercises on each page.

The following are simplified descriptions of the Content standards.

Number and Operations

1. Understands numbers, number representation, relationships, and systems
2. Comprehends operations and their relatedness
3. Fluent in computation and estimation

Algebra

1. Recognizes relationships, functions, and patterns in numbers
2. Uses algebraic symbols
3. Shows similarities in math equations using models
4. Recognizes and evaluates change

Geometry

1. Has knowledge of two- and three-dimensional objects
2. Uses coordinate geometry
3. Recognizes symmetry and transformation
4. Solves problems using geometry

Measurement

1. Understands measurement
2. Uses a variety of methods to show measurement

Data Analysis and Probability

1. Understands the necessity of data
2. Utilizes statistics
3. Makes inferences and predictions based on data
4. Comprehends the basic principles of probability

Table of Contents

Algebra

Geometry

Measurement

Data Analysis and Probability

Problem Solving

Name ____________________

You've Got Value

Write the word value of the underlined digit. The first one is done for you.

1. $3{,}2\underline{7}9$ seven tens	2. $1\underline{7}{,}460$ ________
3. $2\underline{9}0{,}664$ ________	4. $3{,}\underline{8}71$ ________
5. $43{,}82\underline{5}$ ________	6. $16{,}\underline{2}27$ ________
7. $6{,}\underline{4}15$ ________	8. $\underline{1}52{,}428$ ________
9. $5{,}\underline{4}31$ ________	10. $5{,}\underline{6}47$ ________
11. $48{,}5\underline{3}6$ ________	12. $10{,}5\underline{2}9$ ________
13. $\underline{7}9{,}538$ ________	14. $7\underline{2}{,}359$ ________
15. $\underline{6}35$ ________	16. $1\underline{7}4{,}369$ ________
17. $183{,}7\underline{5}2$ ________	18. $3{,}\underline{7}62$ ________

Name ____________________

Ancient Creature

Read each riddle. Find a matching number in the box. Write the number on the line. Then write the letter of each answer below to find the name of an ancient sea creature.

6,513	753	2,406	3,615	2,222	555	1,100
9,598	888	9,703	1,033	7,600	5,287	2,578
642	6,823	9,642	1,191	2,019		

1. I am a 4-digit number. I have a 6 in the hundreds place. My digits add up to 15. What am I? __________ N

2. I am a 3-digit odd number. Each digit is 2 less than the previous digit. What am I? __________ O

3. Each of my digits is odd. They add up to 12. I have 4 digits. What am I? __________ C

4. I am a 3-digit even number. Each digit is the same. What am I? __________ E

5. I am an odd number with a 2 in the tens place. My thousands place is double my ones place. What am I? __________ T

6. I am a 4-digit number greater than 9,599 and less than 9,700. What am I? __________ H

7. I am a 4-digit even number. My digits are arranged from least to greatest. What am I? __________ A

8. If I had 50 less, I would be a 3-digit number. What am I? __________ L

9. Each of my digits is even. One more digit would give me 4 digits. The total of my digits is 12. What am I? __________ C

10. I am an even number. I contain only 2 different numbers. My digits add up to 2. What am I? __________ A

____	____	____	____	____	____	____	____	____	____
642	753	888	1,033	1,100	1,191	2,578	3,615	6,823	9,642

 Name ____________________

That's A Lot

Look at the base-ten blocks. Write the word form, the standard form, and the expanded form. Circle to tell if the number is even or odd.

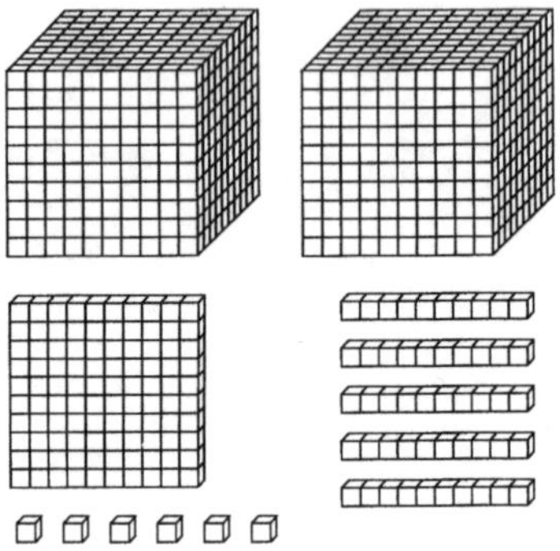

1. word: ______________________________

   standard: ______________________________

   expanded: ______________________________

   circle one: even or odd

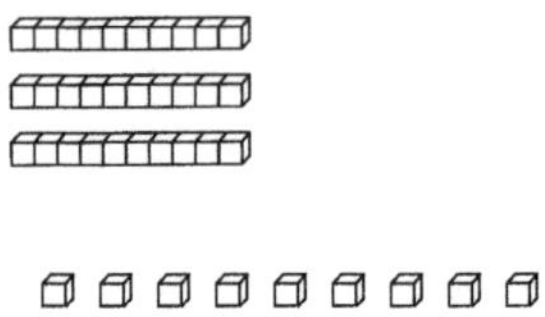

2. word: ______________________________

   standard: ______________________________

   expanded: ______________________________

   circle one: even or odd

Fill in the missing information.

3. word: **three thousand five hundred eighty-two**

   standard: ______________________________

   expanded: ______________________________

   circle one: even or odd

4. word: ______________________________

   standard: **5,137**

   expanded: ______________________________

   circle one: even or odd

 Name ______________________

Standard Form

Write each underlined number in standard form and expanded form.

1. About eight billion four hundred seven candy hearts with messages are made each year.
   standard form: ______________________
   expanded form: ______________________

2. In order for a spacecraft to escape Earth's gravity, it must reach a speed of twenty-five thousand miles per hour.
   standard form: ______________________
   expanded form: ______________________

3. Light travels at about one hundred eighty-six thousand miles per second. It takes light about 8 minutes to travel from the sun to the Earth.
   standard form: ______________________
   expanded form: ______________________

4. The sun is about 92 million miles, or about one hundred forty-seven million kilometers away from the Earth.
   standard form: ______________________
   expanded form: ______________________

5. Halley's comet passes the Earth about every seventy-six years.
   standard form: ______________________
   expanded form: ______________________

6. Sound travels between one thousand one hundred sixteen and 17,100 feet per second.
   standard form: ______________________
   expanded form: ______________________

Number and Operations—comparing number values

Name ____________________

Comparing

When numbers are compared, they can be greater than (>), less than (<), or equal to (=). Place the correct symbol between each set of numbers.

1.	36	______	82	2.	572	______	527
3.	4,627	______	4,627	4.	12,536	______	21,536
5.	74	______	47	6.	362	______	189
7.	4,882	______	4,890	8.	52,218	______	78,000
9.	7	______	11	10.	921	______	912
11.	5,019	______	5,009	12.	37,151	______	15,137
13.	77	______	77	14.	298	______	309
15.	4,657	______	4,567	16.	28,293	______	16,888
17.	23	______	72	18.	575	______	755
19.	2,034	______	1,299	20.	40,362	______	40,382

Chart your answers. Make one tally for each time you used a symbol. Match your tally chart to the correct frequency table below. Circle it.

>	<	=

a.

>	15
<	15
=	10

b.

>	9
<	9
=	2

c.

>	29
<	20
=	1

d.

>	20
<	18
=	2

Number and Operations—rounding whole numbers

Name ____________________

Rounding

Round each number to the nearest ten. Write each number in the correct place in the chart according to whether you rounded up or down.

1. 26 ____________
2. 99 ____________
3. 75 ____________
4. 45 ____________
5. 84 ____________
6. 61 ____________
7. 67 ____________
8. 33 ____________
9. 23 ____________
10. 21 ____________
11. 58 ____________
12. 98 ____________
13. 43 ____________
14. 74 ____________
15. 42 ____________
16. 36 ____________
17. 49 ____________
18. 2 ____________

Round Down	Round Up

Circle each ones digit in the rounding chart. Look at the data. Write the rule for rounding numbers.

____

____

____

____

____

____

____

____

____

Number and Operations—Roman numerals Name ____________

When in Rome

Look at the Roman numerals in each fact. Rewrite as standard numerals.

Basic Roman Numerals

I =	1	V =	5
X =	10	L =	50
C =	100	D =	500
M =	1,000		

1. A manta ray may measure up to XXIX feet wide.

   ____

2. The cheetah is the fastest land animal and can run up to LXX miles per hour.

   ____

3. Candy hearts with messages on them were first made in MCMII.

   ____

4. Decimals were used by Arabs in MCD B.C.

   ____

5. The first American money was made in MDCCXCII.

   ____

6. The calendar we use today was first used by England in MDCCLII.

   ____

7. The XVI-legged monarch caterpillar turns into a VI-legged orange, black, and white butterfly when it emerges from its chrysalis.

   ____

8. "The Star-Spangled Banner" was written by Francis Scott Key in MDCCCXIV but did not become our national anthem until Congress declared it in MCMXXXI.

   ____

9. Sammy Sosa and Mark McGwire both broke baseball's homerun record in MCMXCVIII. Sammy Sosa had LXVI home runs. Mark McGwire had LXX home runs.

   ____

Number and Operations—rounding in base-ten Name ____________________

?  ? Round It ?  ?

Solve. Round each answer to the nearest ten and the nearest hundred.

1.	2.	3.	4.	5.
364	467	572	183	512
+ 298	+ 298	+ 159	+ 457	− 126
r10 ______	r10 ______	r10 ______	r10 ______	r10 ______
r100 ______	r100 ______	r100 ______	r100 ______	r100 ______

6.	7.	8.	9.	10.
900	700	703	279	643
− 317	− 273	− 198	+ 354	− 285
r10 ______	r10 ______	r10 ______	r10 ______	r10 ______
r100 ______	r100 ______	r100 ______	r100 ______	r100 ______

11.	12.	13.	14.	15.
183	502	760	607	573
+ 359	− 131	− 532	+ 296	+ 198
r10 ______	r10 ______	r10 ______	r10 ______	r10 ______
r100 ______	r100 ______	r100 ______	r100 ______	r100 ______

Number and Operations—estimating with whole numbers   Name ______________

Estimate v. Actual

Round each number to the nearest ten.

1. 45 ______   2. 87 ______   3. 264 ______   4. 571 ______

5. 12 ______   6. 36 ______   7. 628 ______   8. 159 ______

Round each number to the nearest hundred.

9. 251 ______   10. 782 ______   11. 367 ______   12. 442 ______

13. 638 ______   14. 546 ______   15. 808 ______   16. 139 ______

Solve. Then round each number of the problem to the nearest hundred. Find the estimated answer using your rounded numbers. Compare the actual and estimated answers.

17. 740 − 252   18. 566 + 295   19. 502 − 114   20. 800 − 121

21. 413 − 125   22. 864 + 249   23. 567 + 256   24. 173 + 338

 Name ____________

Back and Forth

Look at each set of digits. Use the three digits to write the largest possible number on the first line and the smallest possible number on the second line. Write and solve one addition and one subtraction problem with the two numbers.

1. 3, 7, 6
   largest = ______
   smallest = ______
   ____
   ____

2. 5, 3, 8
   largest = ______
   smallest = ______
   ____
   ____

3. 3, 7, 1
   largest = ______
   smallest = ______
   ____
   ____

4. 2, 9, 7
   largest = ______
   smallest = ______
   ____
   ____

5. 7, 1, 4
   largest = ______
   smallest = ______
   ____
   ____

6. 9, 5, 9
   largest = ______
   smallest = ______
   ____
   ____

7. 1, 1, 5
   largest = ______
   smallest = ______
   ____
   ____

8. 8, 2, 5
   largest = ______
   smallest = ______
   ____
   ____

9. 5, 6, 1
   largest = ______
   smallest = ______
   ____
   ____

Name ____________________

Kaleidoscope

Write each missing number. Color each section with an even number yellow. Color each section with an odd number green.

1.

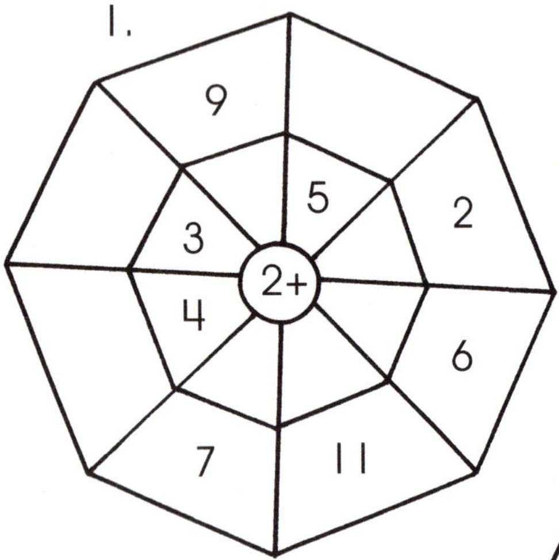

2.

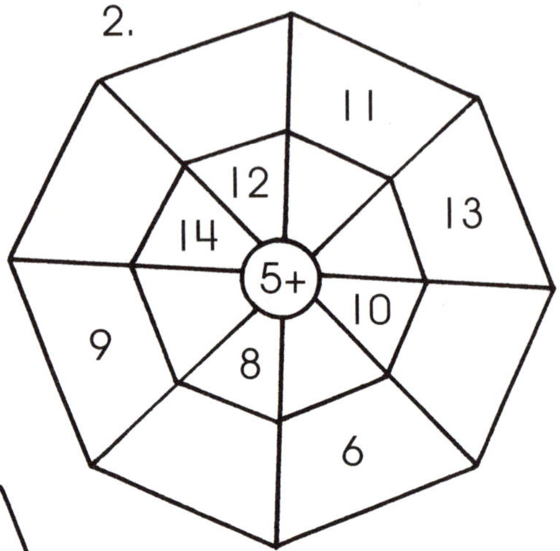

3.

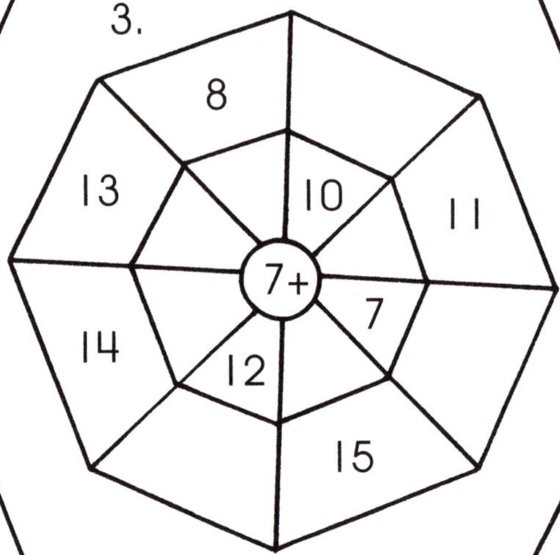

4.

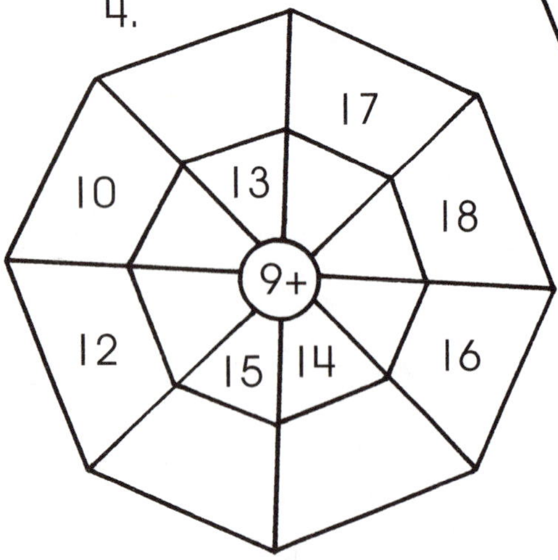

5.

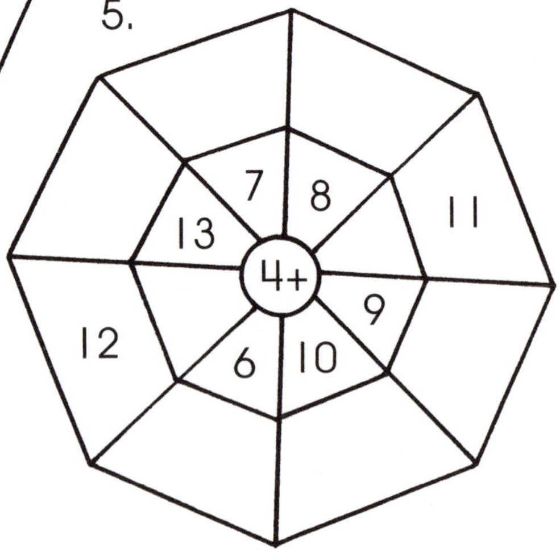

Evaluate your kaleidoscope work. Complete the equations below.

6. even + even = ____________

7 even + odd = ____________

8. odd + even = ____________

9. odd + odd = ____________

Addition Boxes

Place value boxes work well for addition problems with regrouping. Make a box below the equal sign divided into place values. Put a diagonal slash in each box. Add along the diagonal. If there is a remainder, carry it to the next box.

2-digit:

$$\begin{array}{r} 48 \\ +\,59 \\ \hline \end{array}$$

107

3-digit:

$$\begin{array}{r} 497 \\ +\,286 \\ \hline \end{array}$$

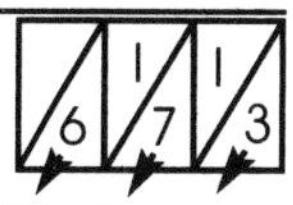

783

Use addition boxes to solve each problem.

1. $48 + 75$
2. $96 + 25$
3. $34 + 58$
4. $17 + 39$
5. $66 + 47$
6. $59 + 21$
7. $45 + 28$
8. $72 + 39$
9. $284 + 486$
10. $397 + 428$
11. $428 + 145$
12. $517 + 387$
13. $259 + 273$
14. $676 + 148$
15. $394 + 429$

Number and Operations—4-digit subtraction with regrouping   Name ____________________

American States

Subtract to find the number of square miles in the following states.

1. Connecticut	9,374 − 4,365	2. New Hampshire	9,868 − 564
3. Delaware	5,346 − 3,289	4. New Jersey	9,618 − 1,782
5. Hawaii	8,396 − 1,946	6. Rhode Island	7,151 − 5,937
7. Massachusetts	9,355 − 1,098	8. Vermont	9,898 − 289

Subtract to find the year the following states were admitted to the U.S.

9. Alabama	8,301 − 6,482	10. Maryland	4,527 − 2,739
11. Alaska	7,277 − 5,318	12. New Mexico	3,384 − 1,472
13. California	5,311 − 3,461	14. Oklahoma	7,834 − 5,927
15. Georgia	3,411 − 1,623	16. Tennessee	9,047 − 7,251

Number and Operations—base-ten and subtraction

Name ____________________

Picture This

Study each base-ten model. Circle the rods and units to be subtracted. Solve for the difference.

1. 64 − 29 = ____

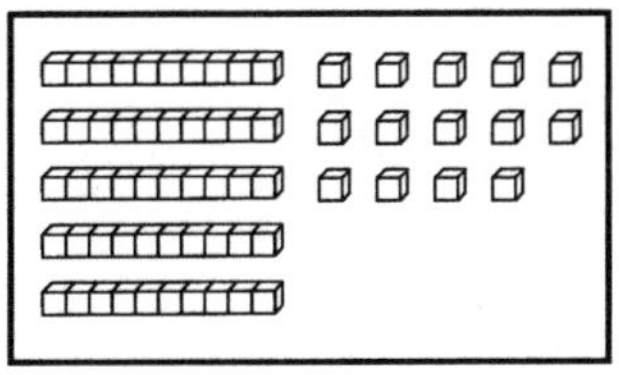

2. 40 − 18 = ____

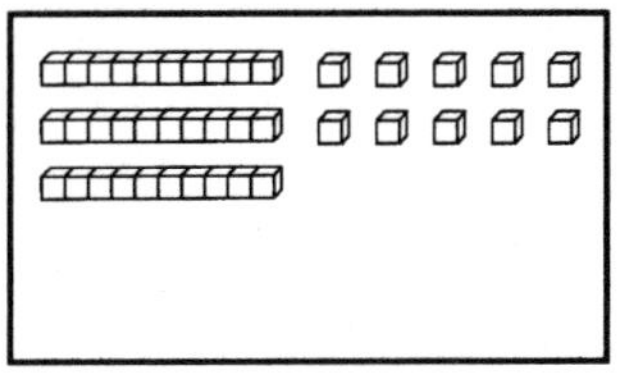

3. 52 − 37 = ____

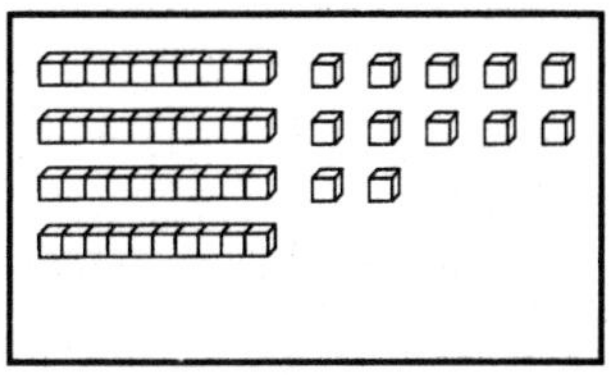

4. 73 − 25 = ____

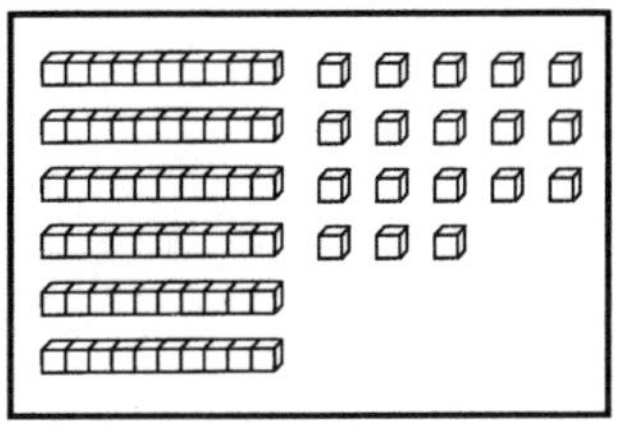

5. 91 − 48 = ____

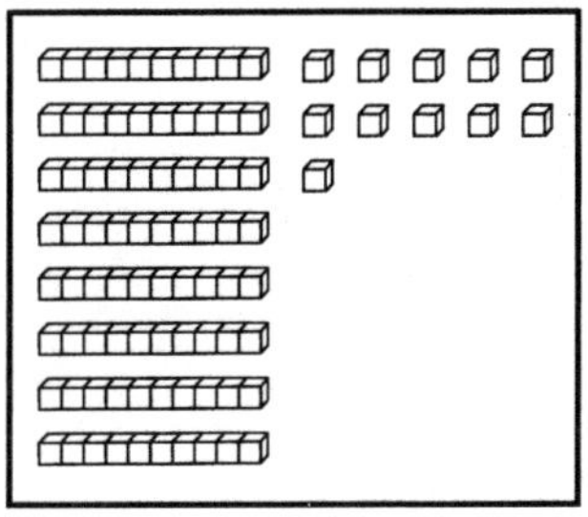

6. 47 − 29 = ____

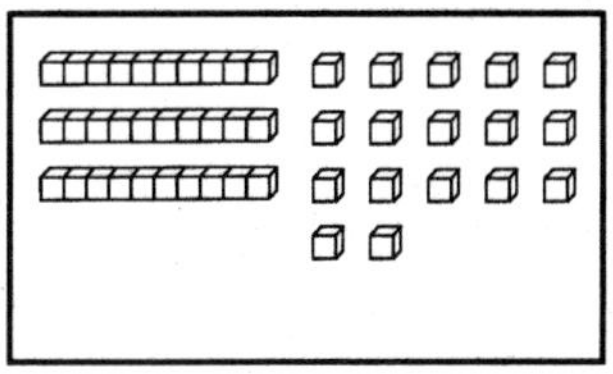

Make a base-ten model for each problem. Solve for the difference.

7. 32 − 17 = ____

8. 90 − 24 = ____

9. 74 − 58 = ____

10. 46 − 39 = ____

11. 81 − 63 = ____

Number and Operations—4-digit addition and subtraction

Name ____________________

Mount Rushmore

Which four presidents are carved on Mount Rushmore in the Black Hills of South Dakota? Underline your predictions. Solve the problems below. If the number after the name is one of the answers, he is on Mount Rushmore.

John Quincy Adams	11,624	Abraham Lincoln	6,823
Grover Cleveland	6,294	James K. Polk	4,011
Ulysses S. Grant	8,663	Theodore Roosevelt	8,117
Thomas Jefferson	9,201	Andrew Jackson	3,709
Lyndon B. Johnson	343	George Washington	7,803
John F. Kennedy	16,458		

1. 4,527 + 516
2. 2,517 + 3,618
3. 1,824 + 6,293
4. 7,251 − 4,524
5. 6,673 − 1,990
6. 2,378 + 5,362
7. 8,461 − 3,163
8. 3,298 + 5,903
9. 4,286 + 3,942
10. 5,201 + 1,622
11. 3,618 + 4,185
12. 6,259 + 1,672
13. 1,374 + 2,843
14. 5,617 + 1,908
15. 9,001 + 1,124

 Name ____________________

Climbing

Find the sums. Use the answers to complete the puzzle and climb the mountain.

Across:

1. 6,431 + 2,986
2. 3,825 + 1,496
3. 7,063 + 1,838
4. 4,957 + 3,479
5. 3,783 + 5,479
6. 1,864 + 1,758
7. 2,439 + 2,588

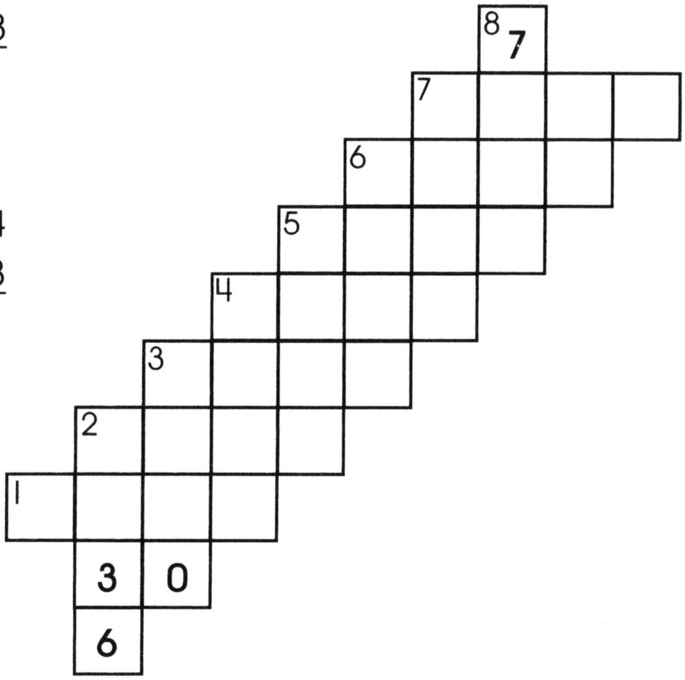

Climb the mountain again by writing the standard numerical form for the numbers reading down. Then write the expanded numerical form.

2. standard: __________ expanded: ____________________
3. standard: __________ expanded: ____________________
4. standard: __________ expanded: ____________________
5. standard: __________ expanded: ____________________
6. standard: __________ expanded: ____________________
7. standard: __________ expanded: ____________________
8. standard: __________ expanded: ____________________

Number and Operations—Roman numerals and addition

Name ______________________

What's Your Score?

Rewrite the problems below in standard numeric form and solve. Shade the Roman numeral answers in the table to find out the percent you got correct.

LXIII	XVII	LXXXVII	CI
XXVIII	XCL	VII	LI
CXIX	CXLII	XXXIV	LIX
CXIV	III	CXXII	XLIV
XVI	XII	LXXV	XXV
XVI	XLIII	CXXX	XXIX

1. XLII − XXVI = ______________________

2. LXXII + LVIII = ______________________

3. XXIX + XXXIV = ______________________

4. LXXVIII − XLIX = ______________________

5. LXV − XXXVII = ______________________

6. XIV + XXXVII = ______________________

7. LIX − XVI = ______________________

8. LV + XLVI = ______________________

9. XCI + XXVIII = ______________________

10. XLIII − XXVII = ______________________

11. LXXX − LXIII = ______________________

12. LXXXI − LVI = ______________________

13. XLIX + XXXVIII = ______________________

14. LXIX + XLV = ______________________

Name ______________________

Rapunzel

Circle the correct symbol between the two equations. To determine which is the real Rapunzel's hair and which is a wig, start at the bottom of the braids and color a section for each correct answer. Color the lowest section on either braid that contains the correct symbol. Which is the real Rapunzel?

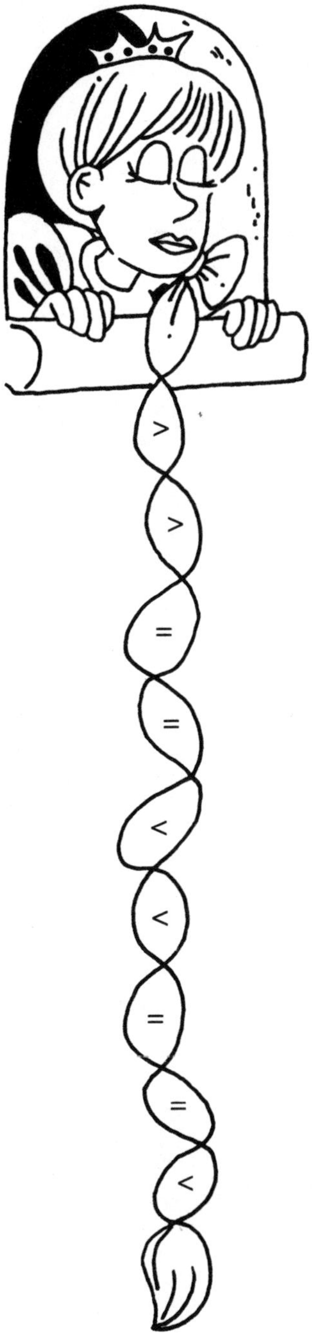

1.	7 + 5	> = <	4 + 9
2.	14 + 6	> = <	9 + 9
3.	10 + 4 + 6	> = <	4 + 9 + 8
4.	15 – 6	> = <	18 – 9
5.	12 – 4 + 7	> = <	18 – 9 + 6
6.	15 – 8	> = <	5 + 1
7.	7 + 6	> = <	14
8.	12 + 6 – 8	> = <	3 + 9 – 0
9.	8 + 6	> = <	5 + 7
10.	5 + 6 + 5	> = <	7 + 8 + 2
11.	6 + 8 + 7	> = <	9 + 7 + 5
12.	4 + 2	> = <	11 – 5
13.	9 + 4	> = <	7 + 7
14.	15	> = <	2 + 6 + 8
15.	13 – 6 + 8	> = <	4 + 6 + 3
16.	11 + 4 – 5	> = <	2 + 8 – 1
17.	1 + 8 + 3	> = <	7 + 6
18.	7 + 9 – 2	> = <	5 – 3 + 8
19.	5 + 4 + 3	> = <	12 – 4 + 5
20.	19 – 9	> = <	3 + 4 + 4

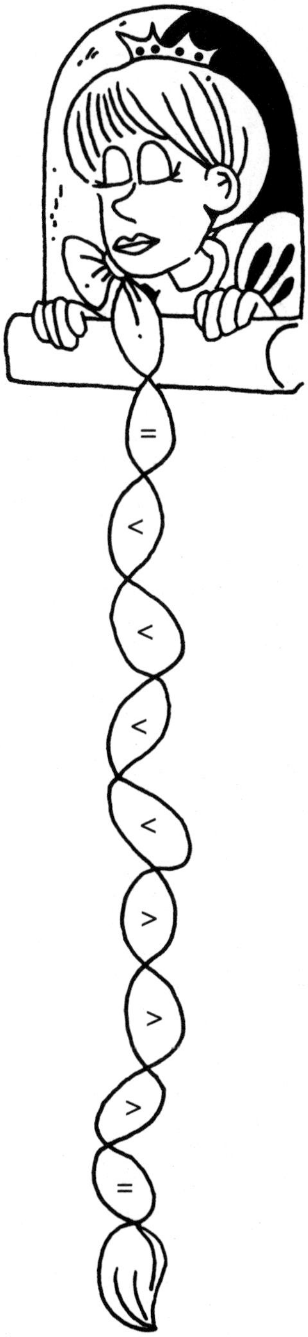

Number and Operations—addition and subtraction

Name ____________________

Table Time

Fill in the addition table below.

+	1	2	3	4	5	6	7	8	9
1									
2									
3									
4									
5									
6									
7									
8									
9									

Solve the subtraction problems to find the differences. Use the addition table to check your answers.

1. 18 − 9 = ______
2. 5 − 2 = ______
3. 12 − 7 = ______
4. 8 − 6 = ______
5. 14 − 5 = ______
6. 13 − 7 = ______
7. 7 − 3 = ______
8. 11 − 6 = ______
9. 14 − 7 = ______
10. 10 − 5 = ______
11. 17 − 8 = ______
12. 17 − 9 = ______
13. 4 − 1 = ______
14. 9 − 3 = ______
15. 13 − 8 = ______
16. 16 − 8 = ______
17. 16 − 9 = ______
18. 14 − 8 = ______
19. 10 − 3 = ______
20. 7 − 2 = ______
21. 12 − 8 = ______
22. 15 − 7 = ______
23. 6 − 5 = ______
24. 12 − 3 = ______

Name ______________________

Order Up!

Solve for each sum. Order the sums from greatest to least in the large box.

1.
$$\begin{array}{r} 0.6 \\ 4.9 \\ +\ 32.5 \\ \hline \end{array}$$

2.
$$\begin{array}{r} 12.60 \\ 7.00 \\ +\ 0.05 \\ \hline \end{array}$$

3.
$$\begin{array}{r} 0.04 \\ 21.20 \\ +\ 6.70 \\ \hline \end{array}$$

4. ______________________
____
____
____

5.
$$\begin{array}{r} 7.90 \\ 0.21 \\ +\ 3.64 \\ \hline \end{array}$$

6.
$$\begin{array}{r} 0.51 \\ 4.06 \\ +\ 0.03 \\ \hline \end{array}$$

7.
$$\begin{array}{r} 3.72 \\ 0.96 \\ +\ 1.70 \\ \hline \end{array}$$

8. ______________________
____
____
____

9.
$$\begin{array}{r} 0.25 \\ 1.70 \\ 0.40 \\ +\ 4.01 \\ \hline \end{array}$$

10.
$$\begin{array}{r} 5.08 \\ 0.16 \\ 2.20 \\ +\ 0.05 \\ \hline \end{array}$$

11.
$$\begin{array}{r} 2.06 \\ 0.62 \\ 1.90 \\ +\ 6.70 \\ \hline \end{array}$$

12. ______________________
____
____
____

 Name ____________________

Sssnakes

Solve each problem. Shade the answers on the snakes below to find out which one slithered to safety.

1. 426.40 − 139.65
2. 346.08 − 263.14
3. 632.12 − 49.60
4. 490.80 − 326.04
5. 104.20 − 63.89
6. 46.00 − 1.34
7. 3.67 − 1.80
8. 246.03 − 158.10
9. 536.48 − 357.00
10. 183.00 − 2.49
11. 826.15 − 683.47
12. 57.40 − 9.63
13. 720.20 − 463.09
14. 100.45 − 34.61
15. 4.63 − 0.99

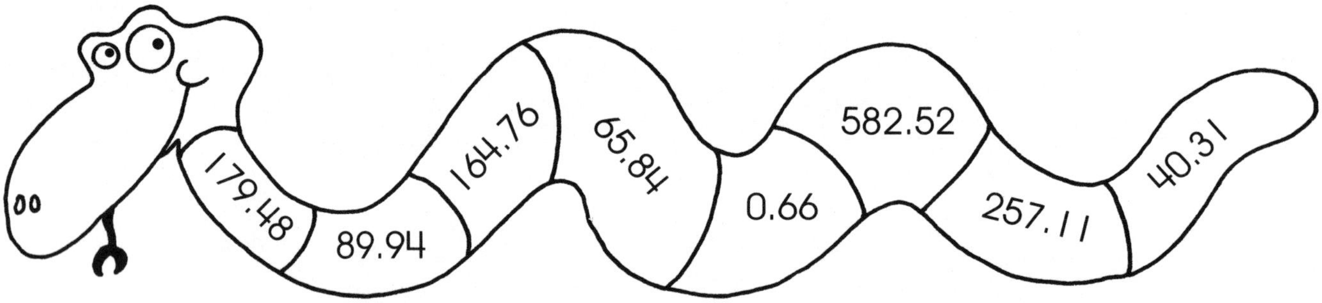

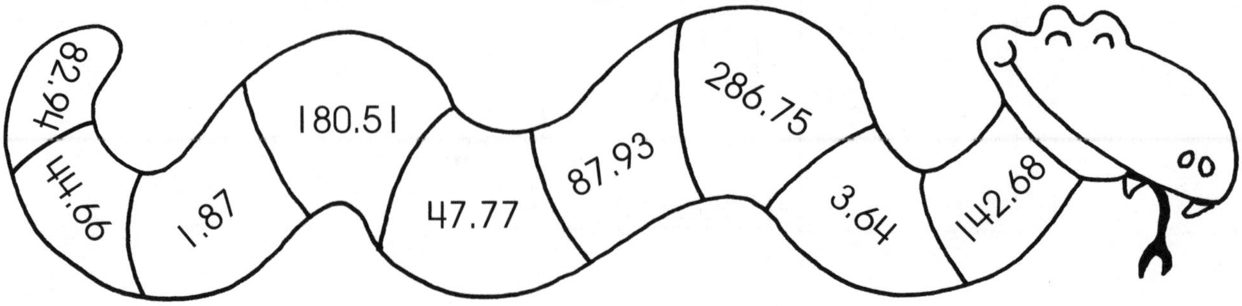

 Name ____________

Multiplication

Draw the sets. Write the multiplication problem each drawing shows.

1. Draw 3 rockets with 5 windows on each. ___ x ___ = ___

2. Draw 5 bears with 2 eyes on each. ___ x ___ = ___

3. Draw 4 pockets with 8 marbles in each. ___ x ___ = ___

4. Draw 2 fish bowls with 6 fish in each. ___ x ___ = ___

5. Draw 5 octopuses with 8 arms on each. ___ x ___ = ___

6. Draw 1 robot with 9 circular parts on it. ___ x ___ = ___

Name ____________________

Traveling

Solve. Draw a line between each product and the corresponding state name in which the ones digits match. Find out in which states the famous places are located.

1.	Acadia National Park	67 x 7	a.	Michigan	82 x 9
2.	Hot Springs National Park	57 x 5	b.	Alaska	42 x 3
3.	John F. Kennedy Space Center	97 x 9	c.	Maine	51 x 9
4.	The Dells	39 x 9	d.	Illinois	42 x 6
5.	Mackinaw Bridge	56 x 8	e.	Florida	27 x 9
6.	Denali National Park	28 x 7	f.	Wisconsin	43 x 7
7.	Sequoia National Park	36 x 5	g.	Oregon	11 x 7
8.	Kilauea Caldera	98 x 3	h.	Arkansas	75 x 3
9.	The Field Museum	73 x 4	i.	California	25 x 6
10.	Crater Lake National Park	19 x 3	j.	Hawaii	32 x 7

Name ______________________

Cobwebs

Write each missing factor. The product is on the outside ring. Color sections of each web with even numbers yellow. Color sections of each web with odd numbers green.

1.

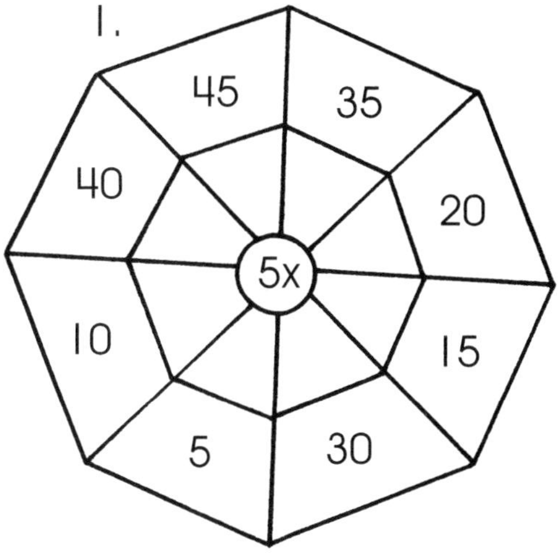

2.

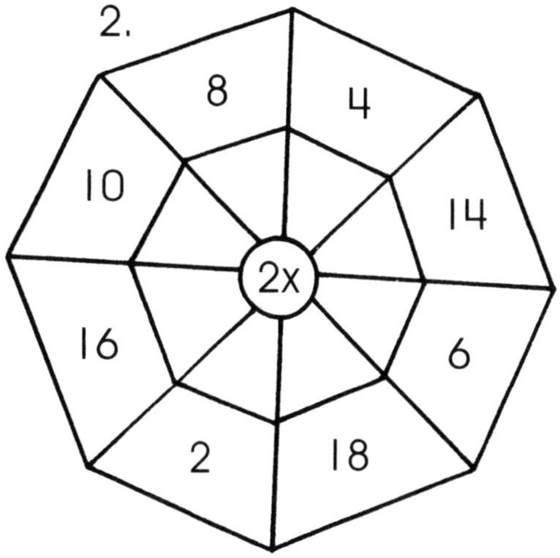

3.

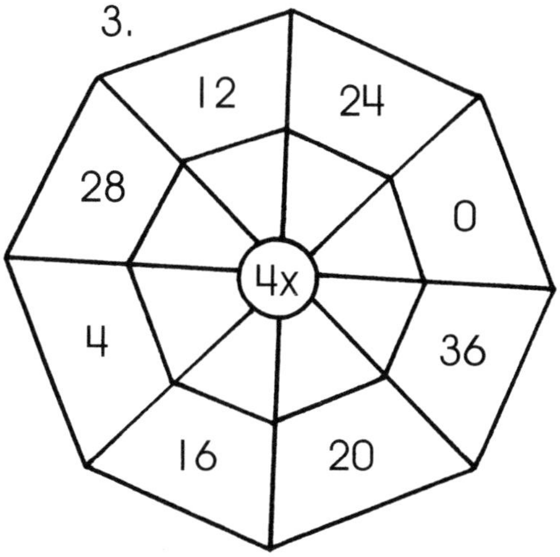

4.

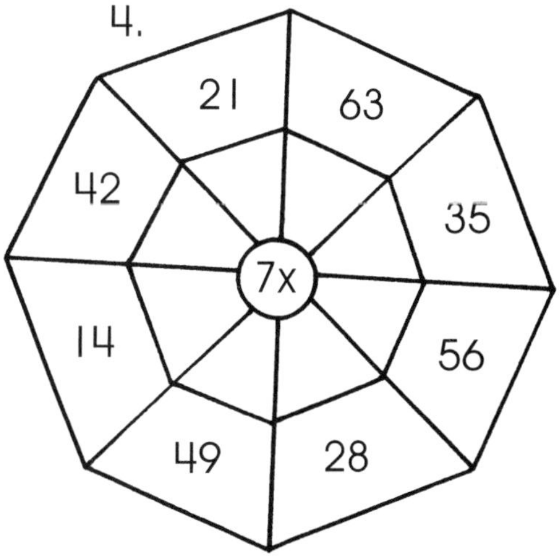

5.

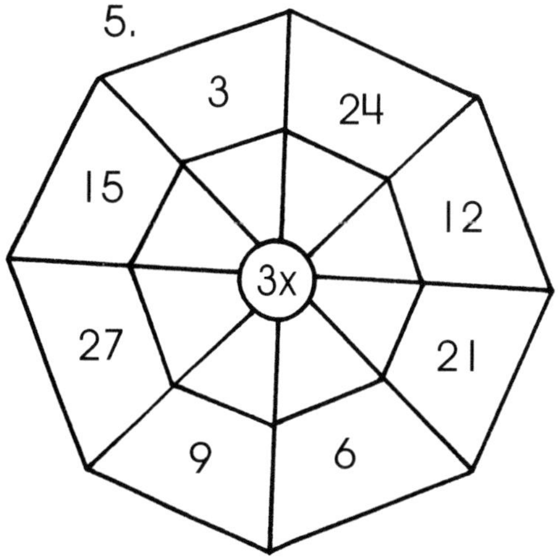

Number and Operations—base-ten and multiplication

Name ____________________

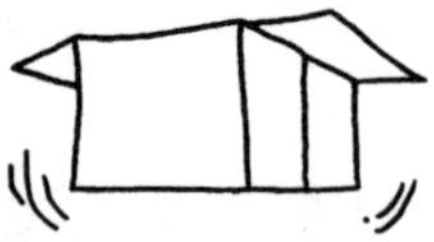

All Boxed Up

Place value boxes work well for multiplication with regrouping. Make a divided box below the equal sign. Put a diagonal slash in each box. To find the answer, add along the diagonal. If there is a remainder, carry it to the next box.

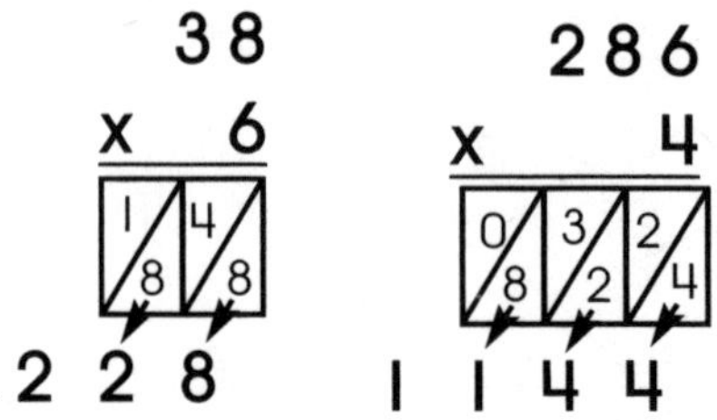

Use place value boxes to solve each problem.

1. 96×6
2. 35×7
3. 527×2
4. 52×7
5. 96×2
6. 219×8
7. 48×9
8. 148×3
9. 722×6
10. 17×3
11. 274×5
12. 382×7

Number and Operations—Roman numerals and multiplication

Name ____________________

For the Romans

Solve each problem. Find the Roman numeral equivalent to the answer in the middle column. Draw a line from the problem to the Roman numeral. Some will not be used.

I = I V = 5 X = 10 L = 50 C = 100 D = 500

1. 53 x 6

2. 47 x 9

3. 32 x 8

4. 21 x 5

5. 68 x 4

a. CDXXIII
b. DLXXXVIII
c. CXXX
d. CCCXVIII
e. CXVII
f. CCLXXII
g. CLXII
h. CCCLX
i. CV
j. CCCXXX
k. CCLII
l. CCLVI
m. CDXXXIII
n. CCXXXVI
o. DXVIII

6. 74 x 7

7. 39 x 3

8. 18 x 9

9. 45 x 8

10. 98 x 6

Number and Operations—base-ten and multiplication

Name ____________________

Two-Digit Place Value Boxes

A place value box with a 2-digit number on the bottom works differently, but is a good tool.

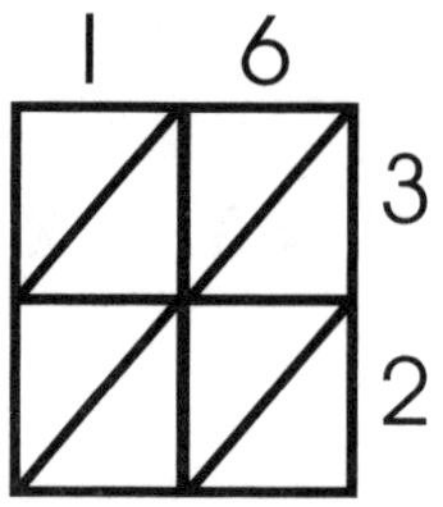

Let's look at the problem: 16 x 32

1. Draw boxes.
2. Write the top factor on top.
3. Write the bottom factor on the side.
4. Place the diagonal slashes in the boxes.
   a. Multiply 2 and 6.
   b. Multiply 2 and 1.
   c. Multiply 3 and 6.
   d. Multiply 3 and 1.

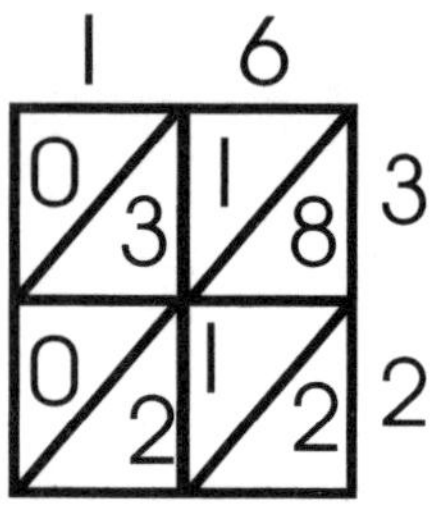

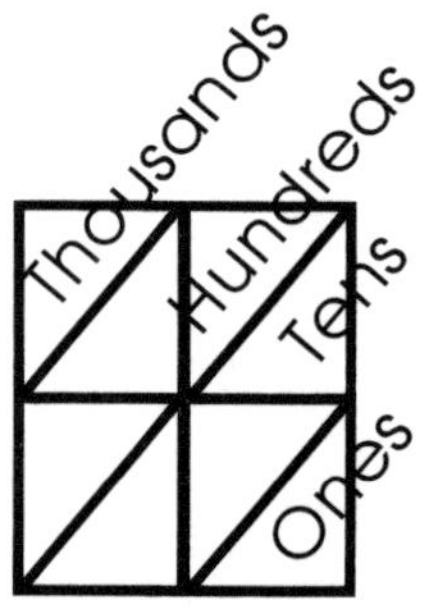

5. Note the place values of the boxes.
6. Add the ones: 2
7. Add the tens: 8 + 2 + 1 = 11 tens (write the one and carry the one into the hundreds place)
8. Add the hundreds: 1 + 3 + 0 + 1 = 5
9. Add the thousands: 0
10. Total of 512.

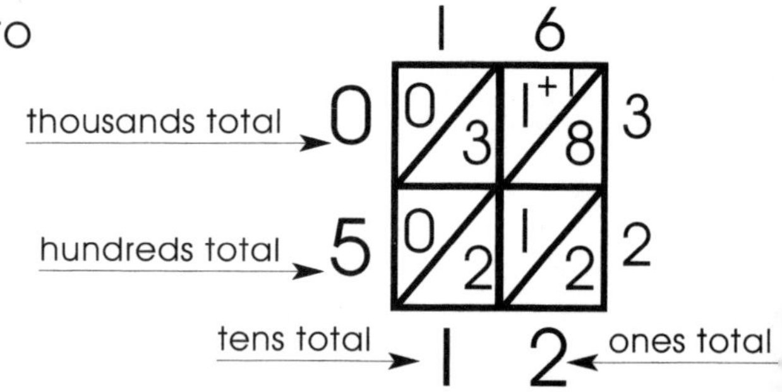

Use the information about place value boxes to help find the products below. You may use the back of this paper if you need more room.

1. 36 x 91 = ____________________
2. 73 x 29 = ____________________
3. 52 x 27 = ____________________
4. 46 x 16 = ____________________
5. 17 x 42 = ____________________
6. 82 x 70 = ____________________
7. 65 x 31 = ____________________
8. 57 x 64 = ____________________

Name ____________________

Multiplying Up

Multiply adjacent numbers together. Write their product in the block above them. Continue multiplying until you reach the top block. Use your skills to fill in the missing blocks.

1.
6	2	5

2.
2	8	3

3.
7	3	9

4.
5	4	3

5.
18		
	2	6

6.
48	32	
	8	

7.
288		
	24	
		4

8.
	45	
3	5	

9.
2	3	1	5

10.
4	3	2	1

11.
640			
20			
	4		
	1		

12.
3	1	2	2	1

13.
5	1	2	1	6

Number and Operations—division

Name ____________________

Division

Draw the sets. Write the division problem each drawing shows.

1. Draw a total of 15 windows in equal groups on 3 rockets.

____ ÷ ____ = ____

2. Draw a total of 10 eyes in equal groups on 5 bears.

____ ÷ ____ = ____

3. Draw a total of 32 marbles in equal groups in 4 pockets.

____ ÷ ____ = ____

4. Draw a total of 12 fish in equal groups in 2 bowls.

____ ÷ ____ = ____

5. Draw a total of 40 arms in equal groups on 5 octopuses.

____ ÷ ____ = ____

6. Draw a total of 9 circular parts in equal groups on 1 robot.

____ ÷ ____ = ____

Number and Operations—multiplication and division

Name ____________________

Table Tool

Fill in the multiplication table below.

x	1	2	3	4	5	6	7	8	9
1									
2									
3									
4									
5									
6									
7									
8									
9									

Solve the division problems to find the quotients. Use the multiplication table to check your answers.

1. 28 ÷ 7 = ______
2. 36 ÷ 6 = ______
3. 63 ÷ 7 = ______
4. 20 ÷ 5 = ______
5. 24 ÷ 4 = ______
6. 8 ÷ 2 = ______
7. 81 ÷ 9 = ______
8. 6 ÷ 3 = ______
9. 54 ÷ 9 = ______
10. 48 ÷ 6 = ______
11. 7 ÷ 1 = ______
12. 27 ÷ 3 = ______
13. 56 ÷ 8 = ______
14. 9 ÷ 9 = ______
15. 54 ÷ 6 = ______
16. 16 ÷ 4 = ______
17. 10 ÷ 2 = ______
18. 64 ÷ 8 = ______
19. 36 ÷ 4 = ______
20. 4 ÷ 2 = ______
21. 35 ÷ 5 = ______
22. 5 ÷ 5 = ______
23. 49 ÷ 7 = ______
24. 12 ÷ 2 = ______

Name ____________________

Schooling

Solve. Add the answers in each small box. In the large box, circle the correct sum of those answers.

1. $3\overline{)81}$	2. $6\overline{)276}$	3. $2\overline{)110}$	4. 130 128 218 134
5. $9\overline{)342}$	6. $4\overline{)276}$	7. $5\overline{)355}$	8. 169 184 182 178
9. $7\overline{)126}$	10. $6\overline{)594}$	11. $8\overline{)192}$	12. 141 137 129 150

 Name ____________________

Practice, Practice

Solve each problem.

1. $6\overline{)78}$
2. $49\overline{)49}$
3. $18\overline{)126}$
4. $9\overline{)117}$
5. $23\overline{)23}$
6. $42\overline{)378}$
7. $7\overline{)133}$
8. $12\overline{)156}$
9. $29\overline{)145}$
10. $11\overline{)132}$
11. $17\overline{)340}$
12. $39\overline{)39}$
13. $31\overline{)124}$
14. $6\overline{)126}$
15. $9\overline{)126}$

Number and Operations—multiplication and division

Name ______________________

Mary Anning

Solve the problems.

1.	21 ÷ 7 = ____	peppers	2.	6 x 4 = ____	a
3.	3 x 4 = ____	of	4.	72 ÷ 8 = ____	peck
5.	14 ÷ 7 = ____	Mary	6.	36 ÷ 6 = ____	by
7.	7 x 7 = ____	picked	8.	56 ÷ 8 = ____	the
9.	3 x 5 = ____	pickled	10.	6 ÷ ____ = 6	seashore
11.	18 ÷ 2 = ____	lamb	12.	6 x ____ = 30	she
13.	2 x 4 = ____	seashells	14.	36 ÷ 9 = ____	sells
15.	4 x 5 = ____	Piper	16.	7 x 4 = ____	had
17.	6 x 6 = ____	Peter	18.	____ ÷ 8 = 4	little

Now solve the following. Circle the tens digit in each answer. Locate the answers above that match the tens digits below. Write the matching words below each problem to find out what tongue twister was inspired by a girl named Mary Anning.

19.	20.	21.	22.	23.	24.
467	601	942	185	526	374
+ 289	− 453	− 553	+ 376	− 248	+ 236

__________ __________ __________ __________ __________ __________

Number and Operations—multiplication and division

Name ____________________

Missing Numbers

Solve to fill in the first empty square. Use that answer to fill in the square where the arrow points. Complete the puzzles.

Puzzle One	Puzzle Two	Puzzle Three
4 x 12 = ☐	3 x 3 = ☐	5 x ☐ = 25
6 x ☐ = ☐	81 ÷ ☐ = ☐	45 ÷ ☐ = ☐
☐ x ☐ = 32	☐ x ☐ = 72	54 ÷ ☐ = ☐
☐ x 9 = ☐	☐ ÷ ☐ = 2	2 x ☐ = ☐
6 x ☐ = ☐	6 x ☐ = ☐	☐ x 4 = ☐
3 x ☐ = ☐	8 x ☐ = ☐	☐ x ☐ = 21
9 x ☐ = ☐	15 ÷ ☐ = ☐	☐ x ☐ = 28
☐ x ☐ = 14	35 ÷ ☐ = ☐	☐ x 5 = ☐
☐ x 4 = ☐	☐ x ☐ = 56	2 x ☐ = ☐
☐ x 2 = ☐	☐ x ☐ = 64	☐ ÷ ☐ = 3

Number and Operations—multiplication and division

Name ____________________

Serving Size

Read each label. Answer the questions.

1. **Sports Drink**
   Total number of servings: ________
   What are the total calories if you drink the whole bottle? ________

Serving size: 8 fl. oz. Calories per serving: 70 Container size: 24 fl. oz.

2. **Skim Milk**
   Total number of servings: ________
   What are the total calories if you drink the whole carton? ________

Serving size: 1 cup (8 fl. oz.) Calories per serving: 90 Container size: 1 pt. (16 fl. oz.)

3. **Peach Tea**
   Total number of servings: ________
   What are the total calories if you drink the whole can? ________

Serving size: 8 fl. oz. Calories per serving: 105 Container size: 1 pt. (16 fl.oz.)

4. **Popcorn**
   Total number of servings: ________
   What are the total calories if you eat the whole bag? ________

Serving size: 1 cup (8 oz.) Calories per serving: 170 Container size: 24 oz.

5. **Canned Peaches**
   Total number of servings: ________
   What are the total calories if you eat the whole can? ________

Serving size: $\frac{1}{2}$ cup (4 oz.) Calories per serving: 106 Container size: 28 oz.

6. **Applesauce**
   Total number of servings: ________
   What are the total calories if you eat 2 cups? ________

Serving size: $\frac{1}{2}$ cup (4 oz.) Calories per serving: 112 Container size: 44 oz.

Name ____________________

Look at the labels on page 40. Find the total number of calories.

7. Raul is having 2 pints of milk, 1 cup of applesauce, and 2 cups of popcorn.

____

8. Safana is having 1 can of peach tea, 1 bag of popcorn, and 1 cup of peaches.

____

9. Jeremy is having 3 pints of milk and 1 bags of popcorn.

____

How many servings of each is he eating? ______________________________

____

10. Maddie is having a 16-ounce glass of sports drink, an 8-ounce glass of milk, $\frac{1}{2}$ cup of applesauce, and $\frac{1}{2}$ cup of popcorn.

____

11. Lupita is having a 16-ounce glass of milk and a bag of popcorn.

____

12. Ian is having 1 can of peach tea and 1 cup of peaches with $\frac{1}{4}$ cup applesauce on top.

____

13. Maegen is having a 16-ounce glass of milk, 8 ounce of peaches, 1 cup of applesauce, and $\frac{1}{4}$ cup popcorn.

____

How many servings of food and drink is she having all together?

____

14. Phan is having half a bag of popcorn and a peach shake. Each shake is made with 16 fluid ounces of milk and 6 ounces of peaches.

____

Name ____________________

Fractions

Use the eight figures to answer the questions below.

figure a

figure b

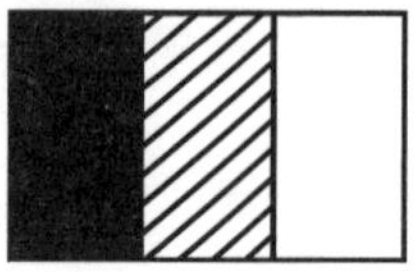

figure c

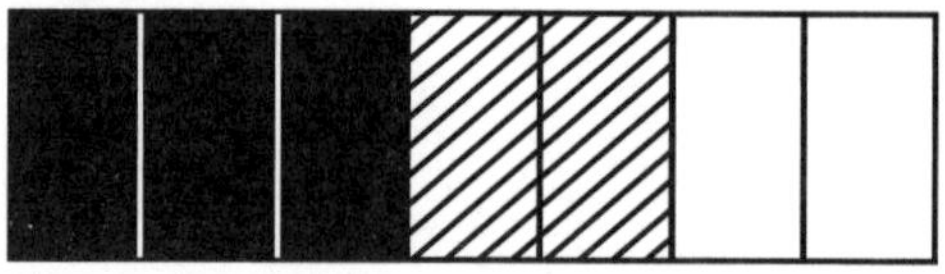

figure d

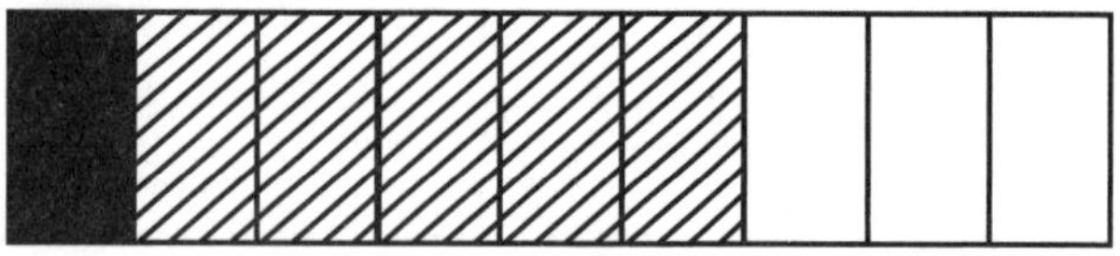

figure e

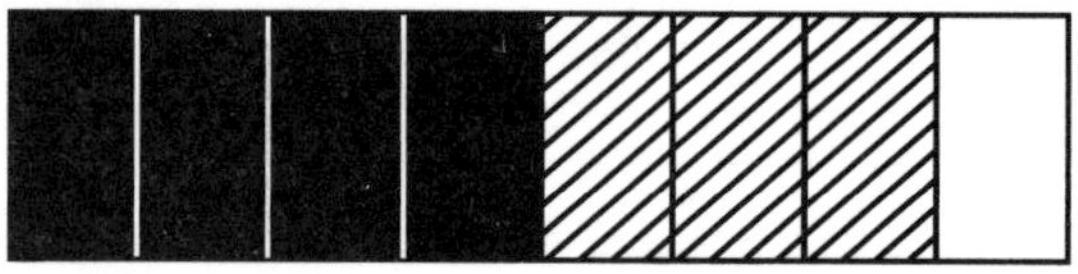

figure f

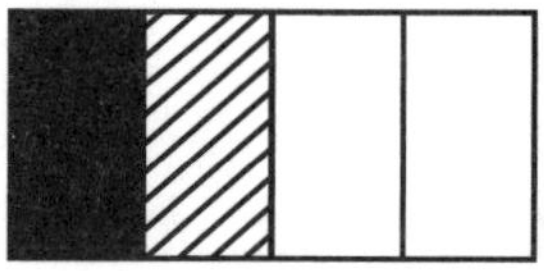

figure g

figure h

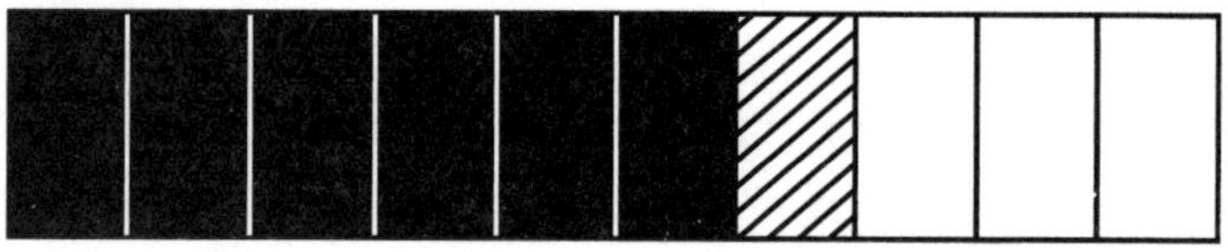

1. Write the fraction for the white part of each figure.
   - figure a ________
   - figure b ________
   - figure c ________
   - figure d ________
   - figure e ________
   - figure f ________
   - figure g ________
   - figure h ________

2. Write the fraction for the black part of each figure.
   - figure a ________
   - figure b ________
   - figure c ________
   - figure d ________
   - figure e ________
   - figure f ________
   - figure g ________
   - figure h ________

3. Write the fraction for the striped part of each figure.
   - figure a ________
   - figure b ________
   - figure c ________
   - figure d ________
   - figure e ________
   - figure f ________
   - figure g ________
   - figure h ________

Number and Operations—fractions as parts of wholes

Name ____________________

Pull Apart

Look at each figure. Write the fraction for the following parts.

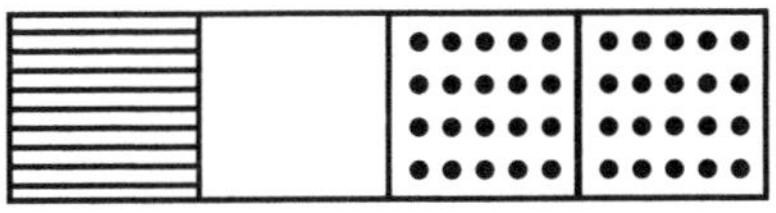

1. The total number of equal parts is ____
   a. white ____________
   b. striped ____________
   c. dotted ____________
   d. white + dotted ____________
   e. white + striped + dotted ____________

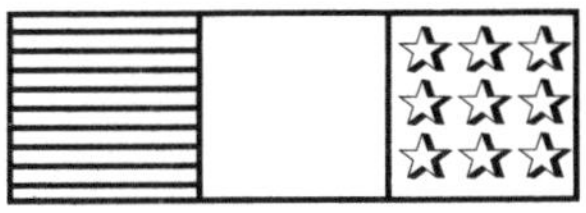

2. The total number of equal parts is ____
   a. striped ____________
   b. white ____________
   c. stars ____________
   d. striped + stars ____________
   e. stars + white ____________

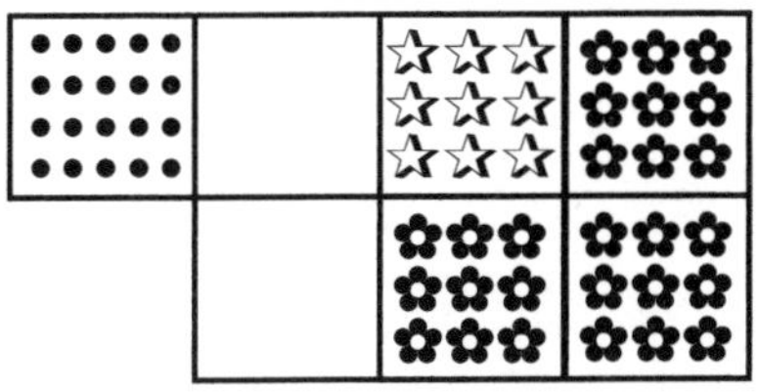

3. The total number of equal parts is ____
   a. white ____________
   b. dotted ____________
   c. flowers ____________
   d. stars ____________
   e. stars + flowers ____________
   f. white + dotted ____________

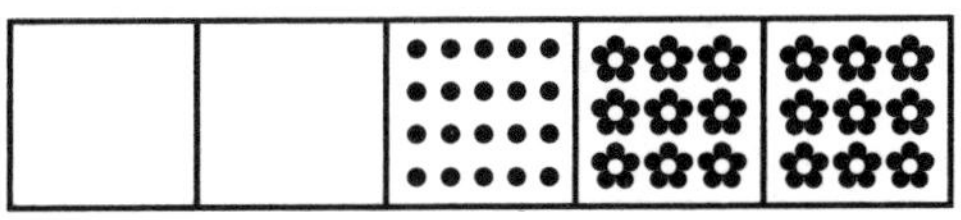

4. The total number of equal parts is ____
   a. dots ____________
   b. white ____________
   c. flowers ____________
   d. white + dots ____________
   e. flowers + white ____________
   f. white + dots + flowers ____________

Name ____________________

Drawing Fractions

Draw each fraction two ways—as part of a whole and as part of a group.

1. $\frac{1}{2}$ is striped.
2. $\frac{2}{7}$ is shaded.
3. $\frac{3}{4}$ is dotted.
4. $\frac{1}{3}$ is striped.
5. $\frac{2}{3}$ is white.
6. $\frac{3}{4}$ is white and starred.
7. $\frac{5}{8}$ is shaded.
8. $\frac{4}{5}$ is dotted and striped.

Challenge: Color this figure so that the following fractions are true.

a. orange + blue = $\frac{4}{9}$

b. red + blue = $\frac{3}{9}$

c. white + yellow = $\frac{4}{9}$

d. yellow + red = $\frac{2}{9}$

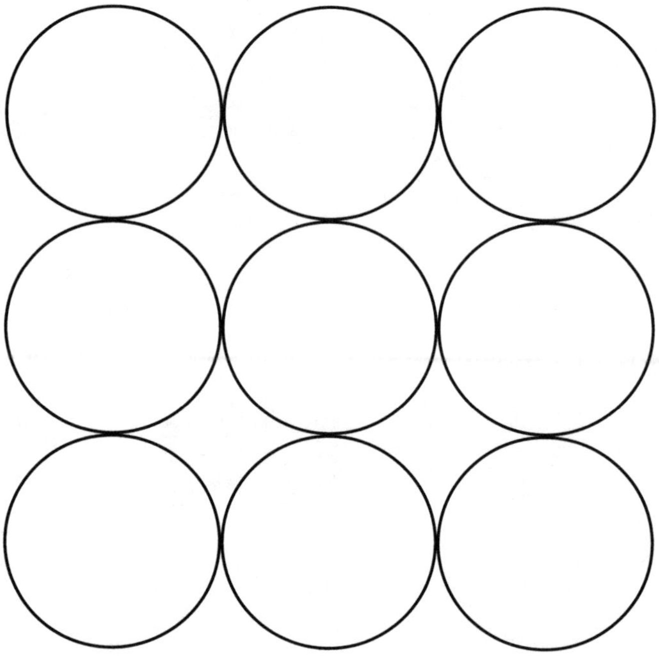

Number and Operations—fractions as parts of sets

Name ______________________

Going Fishing

Color the fish according to the directions.

Color 4 red.
Color 3 blue.
Color 1 yellow.
Color 2 green.
Color 1 orange.
Color 3 purple.

Write the fraction for the following groups of fish.

1. the red fish _____
2. the yellow fish _____
3. the purple fish _____
4. the blue fish _____
5. the orange fish _____
6. the green fish _____
7. the orange and the yellow fish _____
8. the red and the blue fish _____
9. the purple and the green fish _____
10. the green, purple, and red fish _____

11. Circle each way to show $\frac{1}{2}$ of the fish.

    a. red and blue b. orange and purple c. green, orange, and red

12. What is another way to show $\frac{1}{2}$ of the fish? ______________________

____

 Name ____________________

Circles, Squares, and...

Read the directions. Draw the shapes. Color them according to the directions.

1. Draw 3 circles. Color $\frac{2}{3}$ blue.

2. Draw 2 triangles. Color $\frac{1}{2}$ purple.

3. Draw 8 rectangles. Color $\frac{5}{8}$ yellow.

4. Draw 5 ovals. Color $\frac{2}{5}$ green. Color $\frac{3}{5}$ yellow.

5. Draw 4 circles. Color $\frac{1}{4}$ green. Color $\frac{2}{4}$ blue. Color the rest purple.

6. Draw 7 squares. Color $\frac{2}{7}$ brown. Color $\frac{4}{7}$ red. Color the rest white.

7. Draw 4 rectangles. Color $\frac{1}{2}$ red.

8. Draw 8 squares. Color $\frac{3}{4}$ blue.

Name ____________________

Pizza Pie

Look at the fraction beside each pizza. Color each slice that is equivalent to the fraction.

1. 

$\frac{1}{2}$

What fraction of the pie is colored? ____________

2. 

$\frac{2}{3}$

What fraction of the pie is colored? ____________

3. 

$\frac{3}{4}$

What fraction of the pie is colored? ____________

4. 

$\frac{3}{5}$

What fraction of the pie is colored? ____________

In the pizza, write an equivalent fraction in each slice.

5. 

$\frac{1}{3}$

Number and Operations—comparing fractions Name ____________________

Fractionality

Look at the fraction in each square. Color each square using the directions. Then finish the sentence at the bottom of the page.

1. Color spaces blue if they are less than $\frac{1}{2}$. Color them red if they are equal to $\frac{1}{2}$. Color them yellow if they are greater than $\frac{1}{2}$.

$\frac{2}{4}$	$\frac{5}{7}$	$\frac{3}{6}$	$\frac{8}{10}$	$\frac{11}{12}$	$\frac{8}{9}$	$\frac{99}{100}$	$\frac{4}{5}$	$\frac{3}{4}$	$\frac{17}{19}$
$\frac{6}{12}$	$\frac{40}{50}$	$\frac{1}{2}$	$\frac{3}{4}$	$\frac{6}{9}$	$\frac{16}{20}$	$\frac{7}{9}$	$\frac{4}{7}$	$\frac{4}{6}$	$\frac{8}{10}$
$\frac{4}{8}$	$\frac{7}{9}$	$\frac{5}{10}$	$\frac{14}{15}$	$\frac{70}{100}$	$\frac{17}{18}$	$\frac{5}{8}$	$\frac{12}{13}$	$\frac{2}{3}$	$\frac{6}{7}$
$\frac{4}{5}$	$\frac{8}{16}$	$\frac{2}{3}$	$\frac{1}{3}$	$\frac{4}{9}$	$\frac{2}{10}$	$\frac{6}{10}$	$\frac{7}{14}$	$\frac{9}{10}$	$\frac{12}{24}$
$\frac{7}{10}$	$\frac{6}{12}$	$\frac{5}{7}$	$\frac{7}{20}$	$\frac{6}{8}$	$\frac{1}{5}$	$\frac{4}{6}$	$\frac{3}{6}$	$\frac{5}{9}$	$\frac{4}{8}$
$\frac{7}{8}$	$\frac{2}{4}$	$\frac{5}{6}$	$\frac{1}{8}$	$\frac{3}{7}$	$\frac{4}{12}$	$\frac{6}{7}$	$\frac{10}{20}$	$\frac{8}{16}$	$\frac{50}{100}$

2. Color spaces blue if they are equal to 1. Color them red if they are greater than 1. Color them yellow if they are less than 1.

$\frac{4}{5}$	$\frac{3}{6}$	$\frac{6}{10}$	$\frac{3}{3}$	$\frac{2}{7}$	$\frac{5}{4}$	$\frac{3}{6}$	$\frac{4}{9}$	$\frac{2}{5}$	$\frac{2}{2}$
$\frac{3}{7}$	$\frac{1}{13}$	$\frac{1}{9}$	$\frac{10}{10}$	$\frac{1}{3}$	$\frac{7}{11}$	$\frac{3}{5}$	$\frac{3}{10}$	$\frac{1}{6}$	$\frac{4}{4}$
$\frac{1}{2}$	$\frac{7}{7}$	$\frac{2}{2}$	$\frac{5}{5}$	$\frac{2}{6}$	$\frac{16}{10}$	$\frac{3}{7}$	$\frac{6}{6}$	$\frac{3}{3}$	$\frac{11}{11}$
$\frac{11}{12}$	$\frac{6}{6}$	$\frac{1}{9}$	$\frac{8}{8}$	$\frac{1}{7}$	$\frac{8}{3}$	$\frac{1}{5}$	$\frac{7}{7}$	$\frac{4}{7}$	$\frac{35}{35}$
$\frac{7}{9}$	$\frac{1}{1}$	$\frac{6}{6}$	$\frac{4}{4}$	$\frac{3}{4}$	$\frac{16}{7}$	$\frac{4}{6}$	$\frac{9}{9}$	$\frac{16}{16}$	$\frac{5}{5}$

__________ __________ it !

Name ____________________

Capture the Flag

Mikaela and Jade each hid a flag. Follow the directions below. Shade the coordinate of the circled fraction in the odd numbered problems on Mikaela's matrix. Shade the coordinates of the circled fractions in the even numbered problems on Jade's matrix. Who captured a flag?

Circle the smaller of the two fractions.

1. $\frac{1}{2}$ D–5 $\frac{1}{3}$ A–3	2. $\frac{1}{10}$ F–6 $\frac{1}{8}$ F–2	3. $\frac{2}{4}$ C–5 $\frac{3}{4}$ B–6
4. $\frac{1}{6}$ D–5 $\frac{1}{4}$ E–1	5. $\frac{1}{9}$ B–4 $\frac{1}{2}$ C–3	6. $\frac{6}{10}$ C–4 $\frac{9}{10}$ A–1
7. $\frac{1}{7}$ A–1 $\frac{1}{9}$ E–2	8. $\frac{2}{3}$ E–5 $\frac{1}{3}$ A–4	9. $\frac{2}{6}$ D–4 $\frac{4}{6}$ A–4
10. $\frac{1}{4}$ D–2 $\frac{1}{7}$ F–3	11. $\frac{4}{5}$ F–3 $\frac{2}{5}$ C–5	12. $\frac{7}{9}$ D–2 $\frac{1}{9}$ A–3

Circle the larger of the two fractions.

13. $\frac{1}{5}$ F–3 $\frac{1}{3}$ C–6	14. $\frac{1}{10}$ C–1 $\frac{1}{2}$ D–6	15. $\frac{3}{4}$ D–5 $\frac{2}{4}$ F–2
16. $\frac{1}{4}$ B–6 $\frac{1}{2}$ B–4	17. $\frac{1}{4}$ A–5 $\frac{1}{6}$ F–4	18. $\frac{3}{8}$ B–6 $\frac{6}{8}$ C–3
19. $\frac{1}{6}$ D–5 $\frac{1}{3}$ C–4	20. $\frac{1}{5}$ D–5 $\frac{3}{5}$ B–5	21. $\frac{1}{3}$ F–5 $\frac{2}{3}$ E–5
22. $\frac{1}{7}$ E–1 $\frac{1}{8}$ D–6	23. $\frac{7}{9}$ A–4 $\frac{3}{9}$ C–3	24. $\frac{9}{10}$ D–2 $\frac{1}{10}$ F–1

Mikaela

6						
5		F	L	A	G	
4						
3						
2						
1						
	A	B	C	D	E	F

Jade

6						
5						
4		F				
3			L			
2				A		
1					G	
	A	B	C	D	E	F

 Name ____________

Fractional Order

Sketch each fraction in the box above it. Order the fractions from smallest to largest. Read the message by writing the words in that order.

1. $\frac{0}{4}$ this $\frac{4}{4}$ order $\frac{3}{4}$ correct $\frac{2}{4}$ the $\frac{1}{4}$ is

____ ____ ____ ____ ____

2. $\frac{1}{6}$ got $\frac{1}{8}$ you $\frac{1}{2}$ too $\frac{1}{5}$ this $\frac{1}{3}$ one

____ ____ ____ ____ ____

3. $\frac{4}{8}$ way $\frac{1}{8}$ you $\frac{7}{8}$ go $\frac{3}{8}$ bet $\frac{6}{8}$ to

____ ____ ____ ____ ____

4. $\frac{1}{12}$ this $\frac{1}{2}$ just $\frac{1}{16}$ not $\frac{1}{4}$ one $\frac{1}{1}$ kidding

____ ____ ____ ____ ____

5. $\frac{5}{6}$ at $\frac{3}{6}$ are $\frac{6}{6}$ fractions $\frac{4}{6}$ great $\frac{2}{6}$ you

____ ____ ____ ____ ____

 Name ____________

Part Of

Write the fractional part of each number. Find the sum of the fractional parts in each column.

1. $\frac{1}{6}$ of
   a. 24 ________
   b. 18 ________
   c. 12 ________
   d. 36 ________
   e. total ________

2. $\frac{1}{2}$ of
   a. 12 ________
   b. 22 ________
   c. 18 ________
   d. 4 ________
   e. total ________

3. $\frac{1}{3}$ of
   a. 9 ________
   b. 27 ________
   c. 6 ________
   d. 18 ________
   e. total ________

4. $\frac{2}{5}$ of
   a. 15 ________
   b. 40 ________
   c. 25 ________
   d. 10 ________
   e. total ________

5. $\frac{2}{3}$ of
   a. 12 ________
   b. 30 ________
   c. 24 ________
   d. 21 ________
   e. total ________

6. $\frac{4}{7}$ of
   a. 14 ________
   b. 56 ________
   c. 35 ________
   d. 28 ________
   e. total ________

Sal's class has 24 students. The students conducted several surveys about themselves. The results follow. Determine the number of students for each result.

7. $\frac{2}{3}$ are girls.

   ________ students

8. $\frac{7}{8}$ have bicycles.

   ________ students

9. $\frac{4}{6}$ like broccoli.

   ________ students

10. $\frac{1}{2}$ have a brother.

    ________ students

Around the World

Read each clue. Solve each problem. Then use the Code Box to help find the first four letters of each continent.

Code Box
A = 6, 9, 12
E = 3, 14, 19
F = 2, 22, 35
I = 16, 27, 43
N = 13, 20, 25
O = 7, 11, 36
R = 18, 21, 28
S = 4, 10, 30
T = 1, 8, 33
U = 5, 15, 24

1. This continent has the Great Wall and the Yangtze River.

   a. $\frac{4}{9}$ of 27 ______ c. $\frac{3}{5}$ of 45 ______

   b. $\frac{5}{7}$ of 42 ______ d. $\frac{1}{2}$ of 18 ______

   ____

2. The world's largest rainforest and the Amazon River are found on this continent.

   a. $\frac{2}{3}$ of 6 ______ c. $\frac{1}{2}$ of 30 ______

   b. $\frac{1}{4}$ of 44 ______ d. $\frac{2}{5}$ of 20 ______

   ____

3. Ayers Rock is located on this continent that is also a country.

   a. $\frac{1}{8}$ of 48 ______ c. $\frac{2}{7}$ of 14 ______

   b. $\frac{3}{4}$ of 32 ______ d. $\frac{1}{3}$ of 99 ______

   ____

4. This frozen continent is the natural home to penguins, skua, and krill.

   a. $\frac{2}{6}$ of 18 ______ c. $\frac{1}{9}$ of 72 ______

   b. $\frac{4}{5}$ of 25 ______ d. $\frac{3}{7}$ of 21 ______

   ____

Number and Operations—adding fractions Name ______________________

Adding Fractions

Shade the parts to find the sum of each set of fractions.

1. 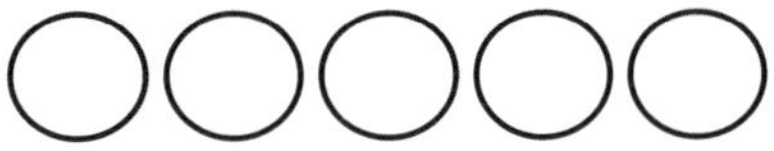 $\frac{2}{5} + \frac{1}{5} =$ ____

2. 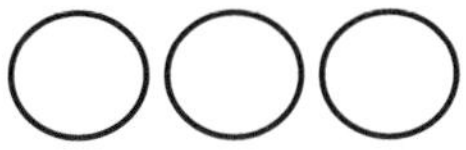 $\frac{1}{3} + \frac{1}{3} =$ ____

3.  $\frac{3}{10} + \frac{5}{10} =$ ____

4.  $\frac{4}{8} + \frac{3}{8} =$ ____

5.  $\frac{1}{12} + \frac{4}{12} =$ ____

6. 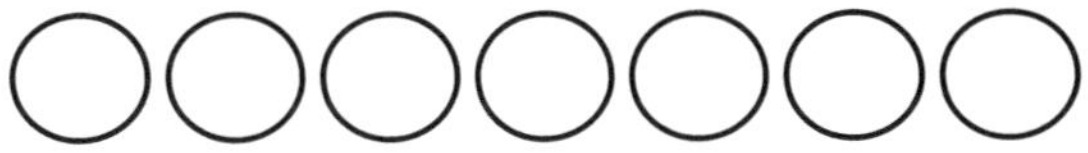 $\frac{4}{7} + \frac{3}{7} =$ ____

Find the sum of each set of fractions.

7. $\frac{3}{7} + \frac{2}{7} =$ ____________

8. $\frac{4}{11} + \frac{5}{11} =$ ____________

9. $\frac{1}{5} + \frac{3}{5} =$ ____________

10. $\frac{7}{12} + \frac{2}{12} =$ ____________

11. $\frac{2}{6} + \frac{2}{6} =$ ____________

12. $\frac{6}{10} + \frac{1}{10} =$ ____________

13. $\frac{2}{8} + \frac{1}{8} =$ ____________

14. $\frac{5}{12} + \frac{6}{12} =$ ____________

 Name ______________________

Subtracting Fractions

Find the difference by crossing out shaded boxes.

1. $\frac{4}{9} - \frac{3}{9} =$ ________

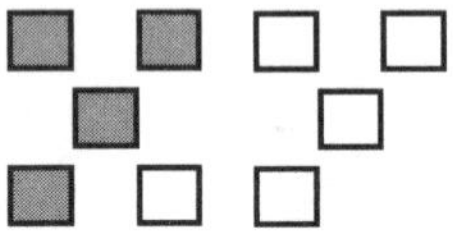

2. $\frac{3}{3} - \frac{1}{3} =$ ________

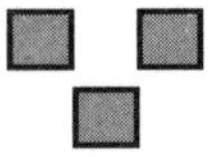

3. $\frac{8}{10} - \frac{1}{10} =$ ________

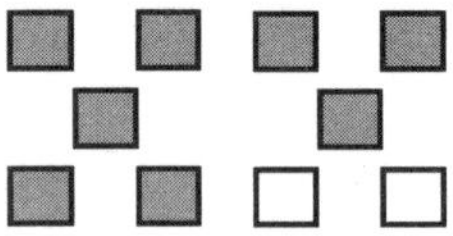

4. $\frac{5}{12} - \frac{4}{12} =$ ________

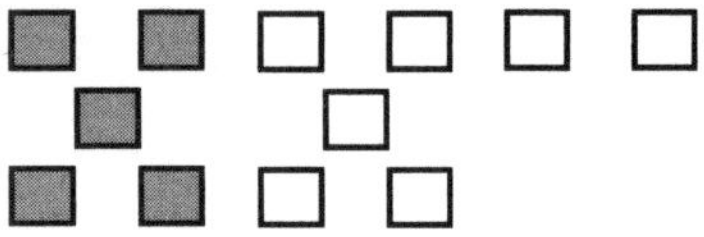

5. $\frac{5}{7} - \frac{3}{7} =$ ________

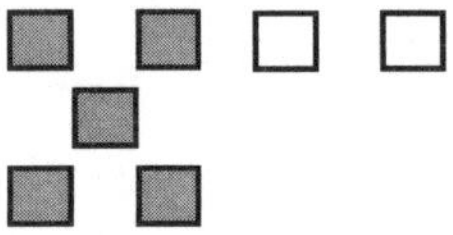

6. $\frac{2}{4} - \frac{1}{4} =$ ________

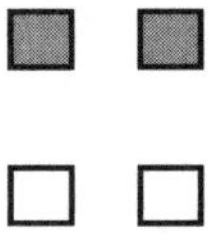

7. $\frac{7}{8} - \frac{3}{8} =$ ________

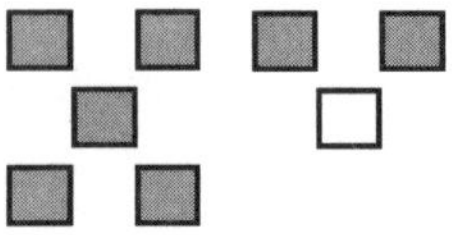

8. $\frac{4}{5} - \frac{1}{5} =$ ________

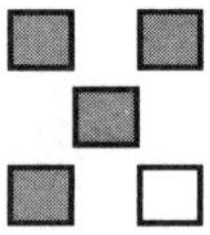

Find the difference of each set of fractions.

9. $\frac{7}{12} - \frac{6}{12} =$ ________________

10. $\frac{4}{7} - \frac{1}{7} =$ ________________

11. $\frac{6}{7} - \frac{2}{7} =$ ________________

12. $\frac{5}{11} - \frac{3}{11} =$ ________________

13. $\frac{3}{8} - \frac{1}{8} =$ ________________

14. $\frac{3}{6} - \frac{2}{6} =$ ________________

15. $\frac{6}{10} - \frac{1}{10} =$ ________________

16. $\frac{5}{8} - \frac{1}{8} =$ ________________

Number and Operations—adding and subtracting fractions

Name ____________________

How Ya Doing?

Solve.

1. $\frac{3}{4} - \frac{2}{4} =$ ______
2. $\frac{4}{5} - \frac{1}{5} =$ ______
3. $\frac{1}{6} + \frac{3}{6} =$ ______
4. $\frac{8}{9} - \frac{6}{9} =$ ______
5. $\frac{2}{10} + \frac{3}{10} =$ ______
6. $\frac{4}{4} - \frac{2}{4} =$ ______
7. $\frac{3}{7} + \frac{3}{7} =$ ______
8. $\frac{2}{5} + \frac{2}{5} =$ ______
9. $\frac{1}{8} + \frac{3}{8} =$ ______
10. $\frac{9}{11} - \frac{4}{11} =$ ______
11. $\frac{6}{7} - \frac{4}{7} =$ ______
12. $\frac{1}{12} + \frac{5}{12} =$ ______
13. $\frac{3}{6} + \frac{2}{6} =$ ______
14. $\frac{10}{11} - \frac{8}{11} =$ ______
15. $\frac{6}{6} - \frac{5}{6} =$ ______
16. $\frac{2}{3} - \frac{1}{3} =$ ______
17. $\frac{7}{9} - \frac{2}{9} =$ ______
18. $\frac{2}{12} + \frac{1}{12} + \frac{2}{12} =$ ______
19. $\frac{1}{7} + \frac{1}{7} + \frac{1}{7} =$ ______
20. $\frac{3}{10} + \frac{1}{10} + \frac{2}{10} =$ ______

Make numbered pairs from the above answers. Use the denominator as the first number and the numerator as the second number. Find the pairs on the graph below and write the letters in order to reveal a message.

1. (__ , __) 2. (__ , __) 3. (__ , __) 4. (__ , __) 5. (__ , __)

6. (__ , __) 7. (__ , __) 8. (__ , __) 9. (__ , __) 10. (__ , __)

11. (__ , __) 12. (__ , __) 13. (__ , __) 14. (__ , __) 15. (__ , __)

16. (__ , __) 17. (__ , __) 18. (__ , __) 19. (__ , __) 20. (__ , __)

6	J	E	P	Z	D	H	D	W	J	B	O	A
5	Y	G	A	R	T	G	I	Y	T	R	N	J
4	O	J	Z	Y	O	U	M	I	W	K	Q	P
3	K	S	P	T	O	A	O	L	D	H	J	X
2	U	Y	C	E	U	Q	G	C	A	R	R	X
1	Z	O	A	Y	B	E	L	O	A	K	J	G
	1	2	3	4	5	6	7	8	9	10	11	12

____

____

____

____

____

____

____

____

Number and Operations—decimals and equivalent forms

Name ____________________

Decimals

Write the decimal that is equal to each word or numeral fraction.

1. eight hundredths

 ____

2. six and twelve hundredths

 ____

3. thirty-one and seven tenths

 ____

4. eleven and seventy-one hundredths

 ____

5. nine and nine hundredths

 ____

6. fifty-two and seven tenths

 ____

7. forty-eight hundredths

 ____

8. three and eighty-one hundredths

 ____

9. $\frac{45}{100}$

 ____

10. $21\frac{3}{10}$

 ____

11. $\frac{56}{100}$

 ____

12. $45\frac{2}{100}$

 ____

13. $98\frac{11}{100}$

 ____

14. $\frac{1}{10}$

 ____

15. $\frac{63}{100}$

 ____

16. $\frac{7}{100}$

 ____

17. $\frac{83}{100}$

 ____

18. $16\frac{5}{10}$

 ____

Number and Operations—decimal, fraction, percent

Name ____________________

Slaying Dragons

To slay a dragon, you must fill in all three parts of the boxes below with equivalent numbers. The first box needs the fraction, the second the percent, and the third needs the decimal. Use the one given to fill in the other two.

	Fraction	Percent	Decimal
1.			0.1
2.	$\frac{23}{100}$		
3.		67%	
4.		5%	
5.	$\frac{40}{100}$		
6.			0.09
7.			0.31
8.	$\frac{2}{100}$		

Name ____________________

Pocket Change

Find the total of each set of coins and write the it on the line. Then complete the word problems.

1. ______________

2. ______________

3. ______________

4. Pia has more than $1.00 but less than $1.30. Write the numbers of the sets of coins that contain the amounts she could have.

   ____

5. If Pia has $1.27, what is the greatest number of dimes she could have? What is the least number of pennies she could have?

   ____

   ____

6. If Pia has at least 1 of each coin, show 1 way she could have $1.27.

   ____

7. Show 2 additional ways she could have $1.27.

   ____

☑ Number and Operations—coin amounts

Name ______________________

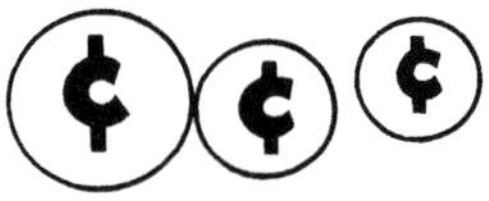

Money Riddles

Read the riddles to determine the coins.

1. I have 5 coins worth 50¢. I have at least 2 different types of coins. What are the coins?

   ____

2. I have 2 different kinds of coins worth 52¢. There are 12 coins total. What are they?

   ____

3. I have 36¢. I have 1 each of 3 different coins. What are my coins?

   ____

4. I have 7 coins worth $1.50. Three is the greatest number of any one coin. What are my coins?

   ____

5. I have 74¢. I have 3 different types of coins. I have an even number of each coin. The coins do not include nickels. What coins do I have?

   ____

6. All 8 of my coins are silver in color. Their total is 50¢. What are they?

   ____

7. I have 5 coins worth 87¢. No nickels are in the bunch. What are the coins?

   ____

8. Show 2 ways to make 68¢. Choose one and write a riddle for it.

   ____

Number and Operations—money problem solving

Name ____________________

Spending Money

Read the directions. Draw the coins. Give the total amount of spending money.

1. Angelica has 6 coins. One-half ($\frac{1}{2}$) are dimes, two-sixths ($\frac{2}{6}$) are pennies, and the rest are nickels. How much does Angelica have?

   ____

2. Manuel has 4 coins. One-fourth ($\frac{1}{4}$) are quarters, one-half ($\frac{1}{2}$) are pennies, and the rest are dimes. How much does Manuel have?

   ____

3. Nell has 12 coins. One-half ($\frac{1}{2}$) are quarters, One-third ($\frac{1}{3}$) are dimes, and the rest are pennies. How much does Nell have?

   ____

4. Judy has 8 coins. One-half ($\frac{1}{2}$) are pennies, one-fourth ($\frac{1}{4}$) are nickels, and the rest are half-dollars. How much does Judy have?

   ____

5. Farley has 10 coins. Three-tenths ($\frac{3}{10}$) are dimes, one-fifth ($\frac{1}{5}$) are quarters, and the rest are half-dollars. How much does Farley have?

   ____

6. Ernie has 9 coins. One-third ($\frac{1}{3}$) are half-dollars, four-ninths ($\frac{4}{9}$) are nickels, and the rest are dimes. How much does Ernie have?

   ____

✓ Number and Operations—money multiplication Name ______________________

Acorns

Five friends are making money by collecting acorns. A local nursery will pay 65¢ a pound for red acorns, $1.25 a pound for black acorns, and 85¢ a pound for white acorns. Write each total and show one way to make that amount in coins.

1. Pedro: 3 lb. black and 6 lb. red ____________

2. Nai: 9 lb. white ____________

3. Faisel: 2 lb. black, 3 lb. red, and 3 lb. white ____________

4. Layne: 4 lb. black, 5 lb. white, and 1 lb. red ____________

5. George: 7 lb. red and 2 lb. white ____________

 Name ____________

Book Order

Use this book order form to answer the questions below.

Beads	$2.75
Discovering Stars	$3.16
Henry Claypot	$4.70
Magic Castle	$2.95
Mastering Multiplication	$2.99
Trolls	$1.86

1. Dorsee bought *Discovering Stars* and *Magic Castle*. What is his total?

   ____

2. Jamal has $5.00. Can he buy *Trolls* and *Magic Castle*?

   ____

3. Anita bought *Beads*. She paid with quarters. How many quarters did she use?

   ____

4. Alan bought *Henry Claypot, Magic Castle,* and *Mastering Multiplication*. He estimated his total would be about $11.00. Compare his estimate to his actual total.

   ____

5. Liza bought *Discovering Stars, Trolls,* and *Henry Claypot*. What is the estimated total? What is the difference between it and the actual total?

   ____

6. Arturo has a 20-dollar bill. He wants one of everything. How much change would he get back?

   ____

Number and Operations—multiplication and division/money

Name ____________________

Post Office

Use mental math to solve the problems. Use the answers to fill in the blanks. Read the story problem and answer the questions.

1. 6 x 4 = ______ 2. 42 ÷ 6 = ____ 3. 7 x 2 = ______ 4. 36 ÷ 6 = ____

Pia and her family of $\underset{4}{\underline{\quad\quad}}$ are going on a $\underset{3}{\underline{\quad\quad}}$-day vacation. They are having the post office hold their mail from June $\underset{1}{\underline{\quad\quad}}$ to July $\underset{2}{\underline{\quad\quad}}$.

5. 48 ÷ 8 = ____ 6. 12 x _____ = 60 7. 63 ÷ 9 = ______ 8. 1 x 1 = ______

Jade needs $\underset{5}{\underline{\quad\quad}}$ stamps. Her sister Mikaela needs $\underset{6}{\underline{\quad\quad}}$, which is one less than Jade needs. How many stamps do they need in all? _______

Each stamp costs $\underset{7}{\underline{\quad\quad}}$¢. What is the total cost of the stamps? ________

Jade has a $\underset{8}{\underline{\quad\quad}}$ dollar bill. How much change will she get back? _____

9. 210 ÷ 7 = ____ 10. 7 x 4 = ______ 11. 8 x 2 = ______ 12. 60 ÷ 30 = ____

13. 2 x 5 = ______

Aaron has to mail some packages. The largest weighs $\underset{9}{\underline{\quad\quad}}$ ounces, the middle-sized package weighs $\underset{10}{\underline{\quad\quad}}$ ounces, and the smallest weighs $\underset{11}{\underline{\quad\quad}}$ ounces. Each package's weight is a multiple of $\underset{12}{\underline{\quad\quad}}$. It costs $\underset{13}{\underline{\quad\quad}}$¢ per ounce to send the packages. What is the total weight of the packages? _________ How much will it cost? __________ Aaron has a $\underset{13}{\underline{\quad\quad}}$ dollar bill. What is his change back? ____________________________

 Name ____________________

Money, Money

Five friends went to the beach. Each brought money for food. Write the amount above each pocket.

1. ____________

2. ____________

3. ____________

4. ____________

5. ____________

Name ____________________

Use the amounts on page 64, the food stand prices below, and the problem-solving matrix to find out how much money each person brought to the beach.

hot dog	$1.37	soda	$0.73	chips	$0.54
burger	$2.18	ice cream	$1.09		

- Sep does not have enough money to buy ice cream for himself and his four friends.
- Ruy has enough to buy 3 hot dogs and 4 sodas.
- If Ann buys 1 burger, 1 soda, 1 bag of chips, and 1 ice cream, she could not get anything else.
- Ziv doesn't have enough for 2 burgers, 1 hot dog, and 1 soda.

	Pocket 1 ______	Pocket 2 ______	Pocket 3 ______	Pocket 4 ______	Pocket 5 ______
Ann					
Eva					
Ruy					
Sep					
Ziv					

How much money does each friend have?

Ann has __________

Eva has __________

Ruy has __________

Sep has __________

Ziv has __________

Name ____________________

Farmer's Market

Ephrain and his class went on a field trip to the Farmer's Market. Their assignment was to find out the cost of produce and determine change back. Help Ephrain complete the total cost and change back columns.

produce	price per pound	pounds bought	total cost	paid with	change back
apples	88¢	47		$42.00	
avocados	79¢	11		$9.00	
blueberries	67¢	37		$25.00	
broccoli	52¢	14		$8.00	
cantaloupe	83¢	15		$13.00	
carrots	27¢	34		$10.00	
cucumbers	36¢	8		$5.00	
green beans	48¢	29		$14.00	
peaches	98¢	21		$21.00	
tomatoes	86¢	12		$10.50	
strawberries	93¢	43		$40.00	

Choose the amount in the change back column for 2 items. Write, sketch, or stamp one way the change could be returned for each.

1.

2.

Name ____________________

Envisioning

Mathematics involves picturing situations. Recognizing a picture pattern helps you notice relationships faster. Look at the numbers pictured below. Write the number of dots under each domino.

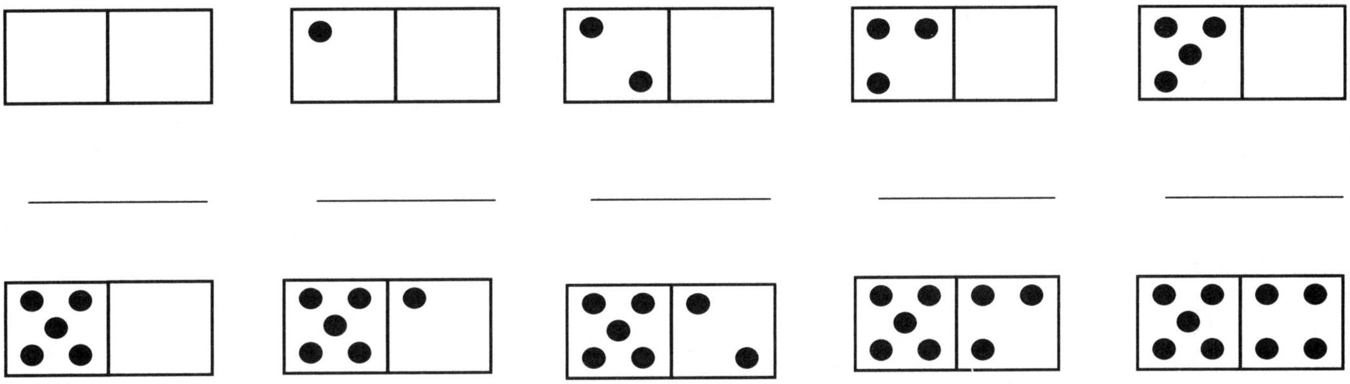

This number pattern can be used when adding, subtracting, multiplying and dividing. Look at the pictured dominoes. Write two number sentences that could be made with the numbers.

1. 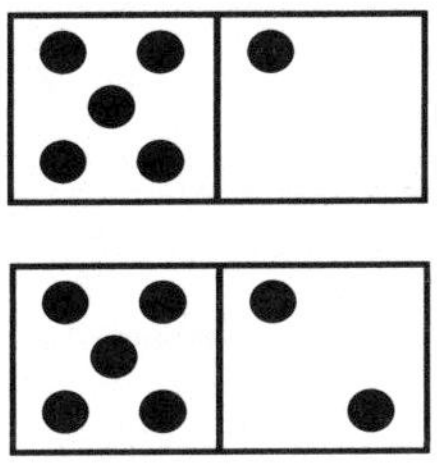

2. 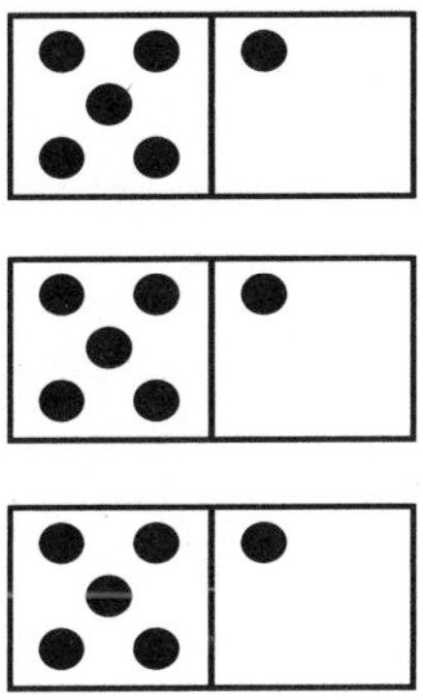

3. 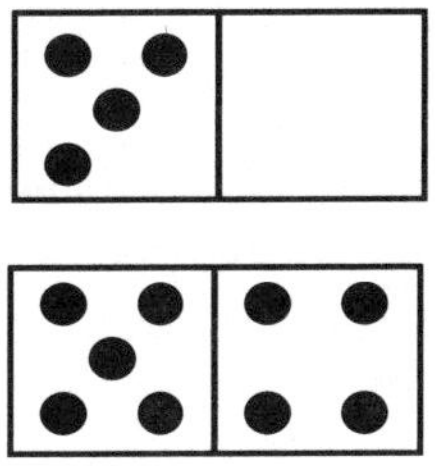

Name ____________________

Left to Right

Look at each series of numbers. Look for patterns and relationships. Group adjacent numbers to make true equations. You may use more than three numbers at a time. Find four true equations for each line.

Example: 2 6 12 4 3 7 0 1 5 3 2 4

$5 = 3 + 2$ $(1 \times 5 + 3) \div 2 = 4$ $12 - 4 < 3 + 7$ $2 \times 6 = 12$

1. 16 8 2 2 4 1 5 10 8 2 9 25

______ ______ ______ ______

2. 10 7 3 21 5 2 9 3 3 7 1 7

______ ______ ______ ______

3. 17 1 2 8 10 3 6 1 8 7 6 5

______ ______ ______ ______

4. 0 15 0 5 3 7 1 4 5 9 0 6

______ ______ ______ ______

5. 6 1 6 3 18 9 9 0 7 1 8 7

______ ______ ______ ______

Name ______________________

Going to the Top

Add adjacent numbers together. Write their sum in the block above them. Continue adding until you reach the top block.

1.

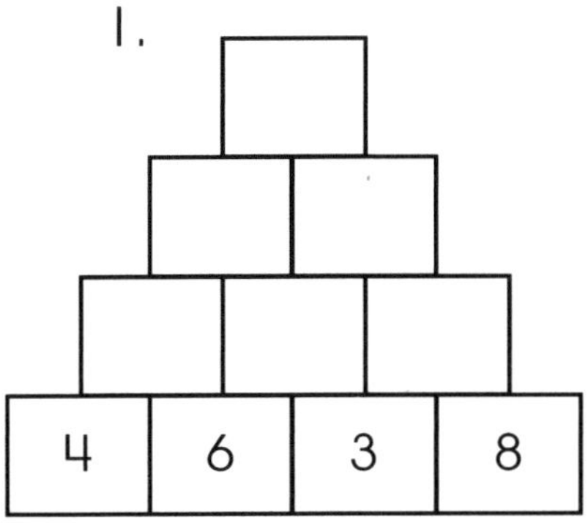

2.

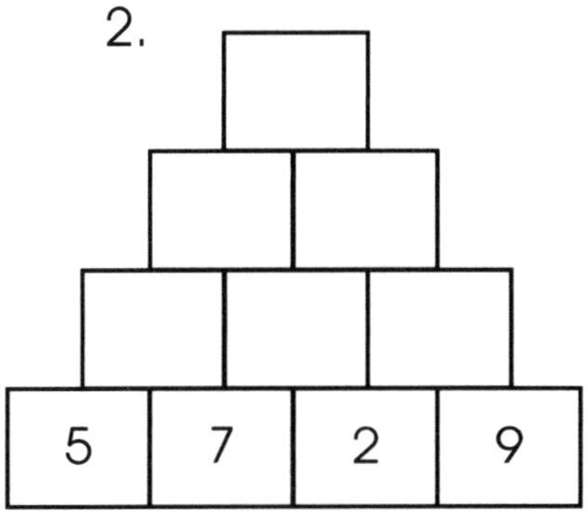

3.

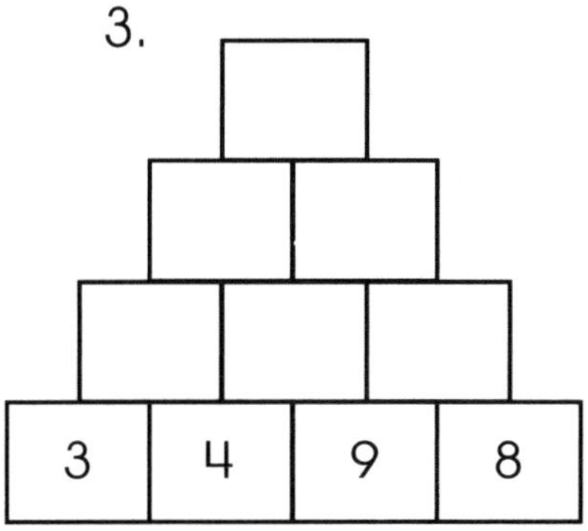

4.

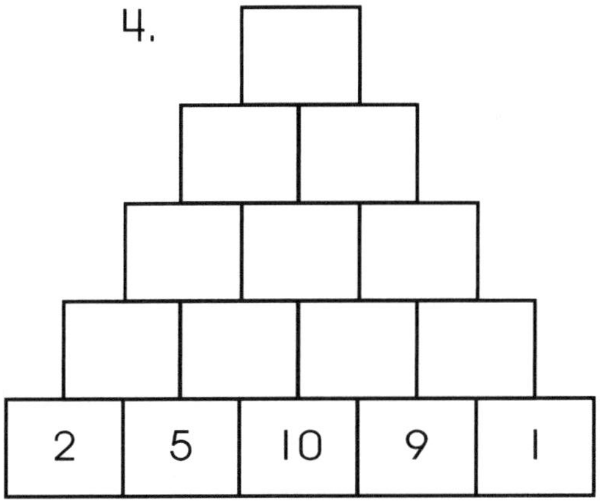

5.

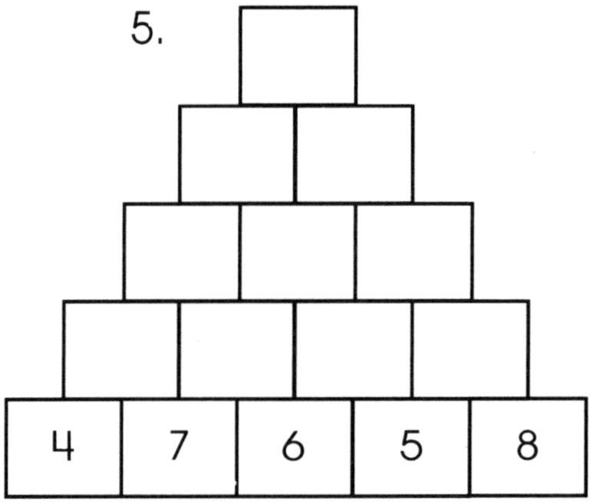

6.

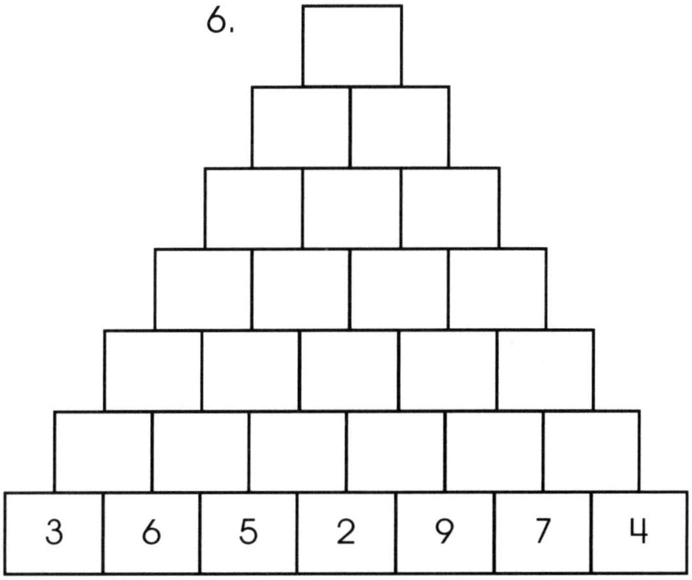

Algebra—unknown quantity Name ____________________

Brick Work

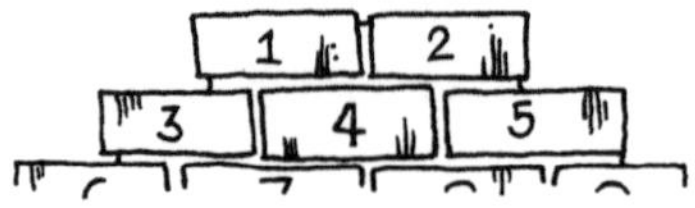

Use your skills to write the missing digits in each problem. Add the missing digits in each row. Write the sum in the circle.

1. 24 + □5 = 59	2. 1□ + 31 = 47	3. 63 + 1□ = □0	4. 46 + □□ = 74	S
5. 46 + □4 = 9□	6. 73 + 2□ = □3	7. □8 + 31 = 7□	8. 20 + 3□ = □7	A
9. 45 + □3 = 88	10. 2□ + 6 = □1	11. □6 + 1□ = 26	12. 31 + 23 = □□	M
13. □3 + 26 = 3□	14. 41 + 4□ = □3	15. □5 + 2□ = 72	16. 64 + □3 = 8□	N
17. □□ + 20 = 65	18. 24 + □□ = 36	19. 27 + □□ = 50	20. 3□ + 16 = □3	O

What is someone called who works with bricks, stone, or cement? Find each sum in the circles and write the letter on the line below.

____ ____ ____ ____ ____

22 38 34 30 40

Algebra—variables as values

Name ____________________

Give Me an N

Use the information given to find the missing numbers. The value of the letters will change with each new problem.

1. $n = 5$
   $n + n + n + n = f$
   $f =$ ____________________

2. $t = 6$
   $t + m = 14$
   $m =$ ____________________

3. $p = 8$
   $p = w + w + w + w$
   $w =$ ____________________

4. $r = 6$
   $r + r + r = h - r$
   $h =$ ____________________

5. $a = 21$
   $21 - c - c = c$
   $c =$ ____________________

6. $k = 8$
   $k - 3 = n$
   $n + k = j$
   $n =$ ____________________
   $j =$ ____________________

7. $z = 7$
   $z + i + i = 13$
   $i = 5 - v$
   $i =$ ____________________
   $v =$ ____________________

8. $p + g = w$
   $y + y = p$
   $w = y + y + y + y + y$
   $p = 4$
   $g =$ ____________________
   $w =$ ____________________
   $y =$ ____________________

9. $f + f = p + d$
   $d + d + d = f$
   $m - d = f + h$
   $p + p = m$
   $d = 1$
   $f =$ ____________________
   $h =$ ____________________
   $m =$ ____________________
   $p =$ ____________________

Algebra—variables with multiplication

Name ____________________

Number Letters

Each problem has a letter in place of a number. Use what you know about multiplication and division to determine the missing number.

1. $3 \times m = 12$

   m = __________

2. $5 \times 8 = t$

   t = __________

3. $b \times 7 = 42$

   b = __________

4. $2 \times d = 4 \times 4$

   d = __________

5. $45 = n \times 5$

   n = __________

6. $6 \times 4 = a \times 8$

   a = __________

7. $u \times 3 = 6 \times 5$

   u = __________

8. $e = 7 \times 3$

   e = __________

9. $24 = r \times 2$

   r = __________

10. $4 \times t = 14 \times 2$

    t = __________

11. $10 \times 4 = s \times 8$

    s = __________

12. $3 \times 6 = 9 \times o$

    o = __________

13. $14 \times e = 2 \times 7$

    e = __________

14. $l = 6 \times 3$

    l = __________

15. $c = 9 \times 3$

    c = __________

Use the letters from each problem above matched with the answers below to reveal a message.

___ ___ ___ ___ ___ ___   ___ ___ ___ ___ ___ ___ ___
5 7 1 3 13 9   14 13 10 10 13 9 11

___ ___ ___ ' ___   ___ ___ ___ ___ ___   ___ ___!
4 12 5 10   11 15 6 9 13   1 13

Name ______________________

Order Up

Read the directions. Sketch and label each line of objects in the order given from left to right.

1. A crayon is two behind a marker. The pencil is in front of the chalk. The marker is three in front of the pen. The pencil is not last.

2. The mouse is next to the cat. The bird is in front of the dog. The cat is third in line. The ant is three behind the mouse.

3. The glass of water is in the middle. The milk is in front of the water and root beer. The root beer is behind the cola. The orange juice is not first, but comes before the cola.

4. The apple is in front of the banana. The pear is behind the banana. The grapes are in front of the orange and banana. The orange is fourth. The apple is not second.

5. The triangle is not first. The quadrilateral is fourth. The triangle is behind the hexagon. The circle is in front of the quadrilateral. The third object has no vertices. The octagon is behind the quadrilateral.

Name ____________________

Scavenger Hunt

Maegen went on a scavenger hunt at Forest Park. Her team needed to find the following objects. Give the coordinates where each object was found.

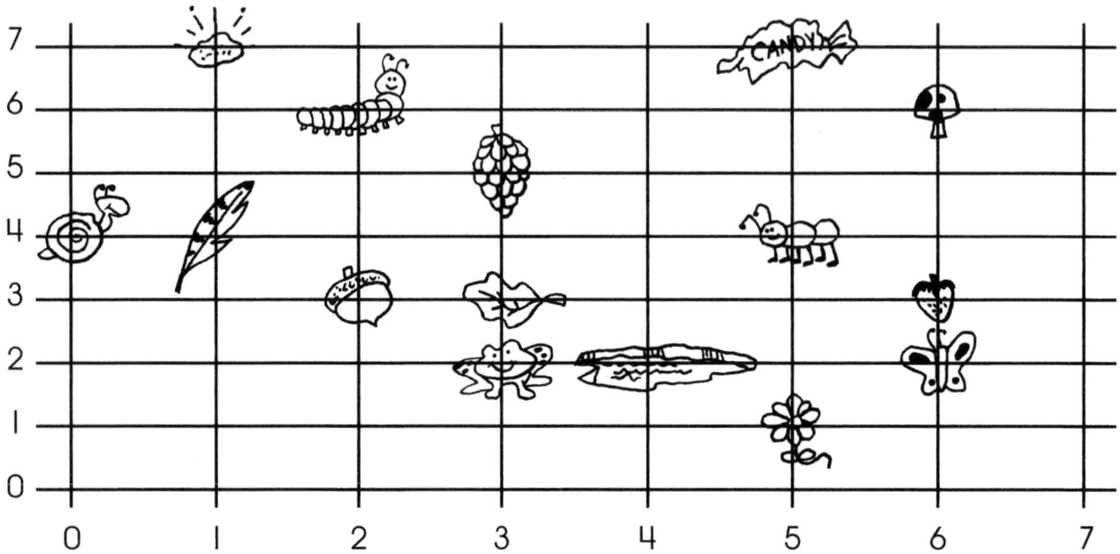

1. __________ acorn
2. __________ leaf
3. __________ ant
4. __________ pinecone
5. __________ butterfly
6. __________ pond
7. __________ candy wrapper
8. __________ snail
9. __________ caterpillar
10. __________ sparkley rock
11. __________ feather
12. __________ mushroom
13. __________ flower
14. __________ strawberry
15. __________ frog

In order to win, the team must also do the following. Help them win by drawing the given shapes at each coordinate.

- Put a birdhouse at (4,6).
- Leave sunflower seeds at (0,2).
- Plant a tree at (7, 5).
- Hang a bat house at (4, 4).
- Put a clean bag in a trash can at (1, 1).
- Clean a water fountain at (7, 1).

Geometry—coordinate graphing Name ____________

Are You Included?

Plot each point on the graph. Connect the (x, y) coordinates.

(1, 5), (1, 3), (3, 1), (5, 1), (7, 3), (7, 5), (5, 7), (3, 7), (1, 5)
Stop and connect the points.

(4, 6), (4, 0), (10, 0), (10, 6), (4, 6)
Stop and connect the points.

(3, 3), (9, 3), (5, 10), (3, 3)
Stop and connect the points.

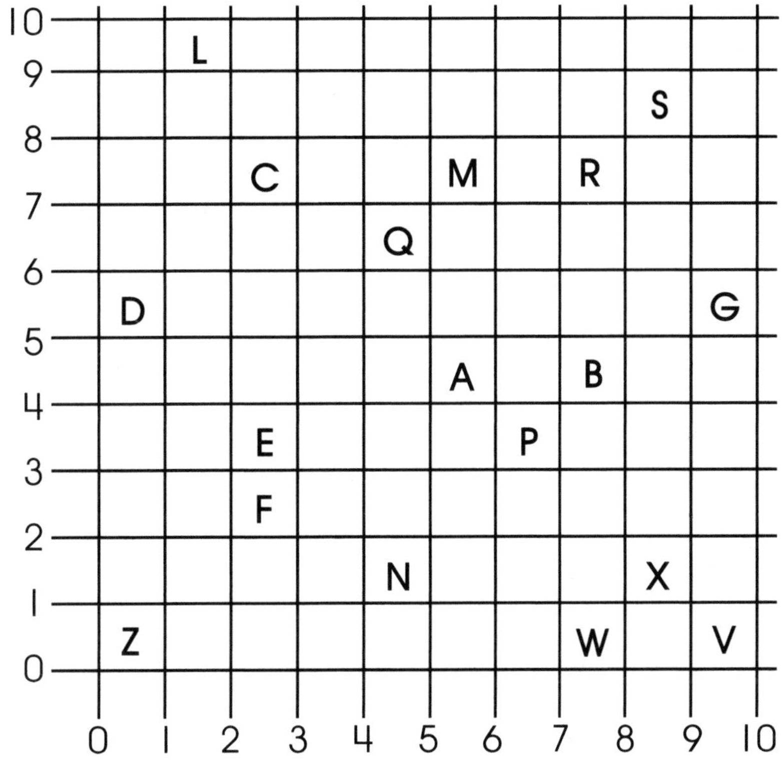

Answer the following questions based on the figures above. Tell which letters are included in the given sets.

1. in the square, but not in the triangle ____________
2. in the intersection of the triangle and the square ____________
3. in the octagon only ____________
4. in the intersection of the triangle and the octagon ____________
5. in the pentagon ____________
6. only in the triangle ____________
7. in the pentagon or octagon, but not in the triangle ____________
8. in the intersection of the pentagon, triangle, octagon, and square ____________
9. not in the pentagon, triangle, octagon, or square ____________

Are We the Same?

Write yes or no to tell if each pair below is congruent.

1. ____________________

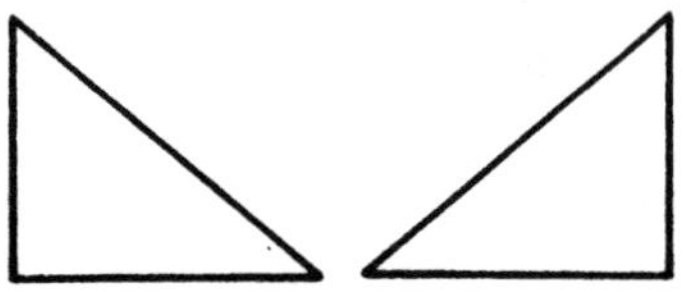

2. ____________________

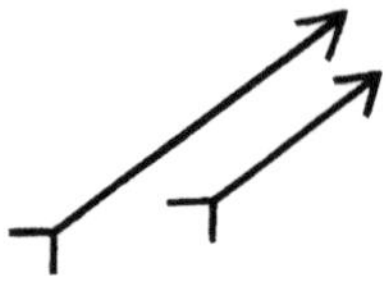

3. ____________________

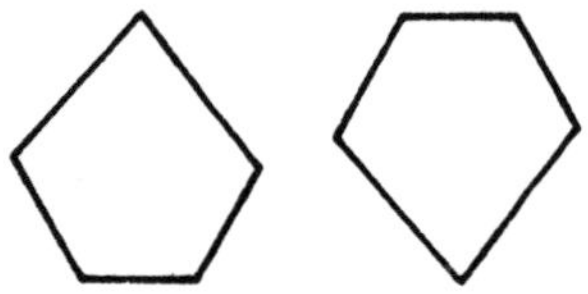

4. ____________________

5. ____________________

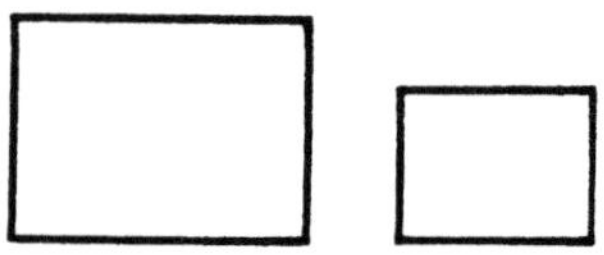

6. ____________________

Look at each shape below. Draw a shape that is similar.

7.

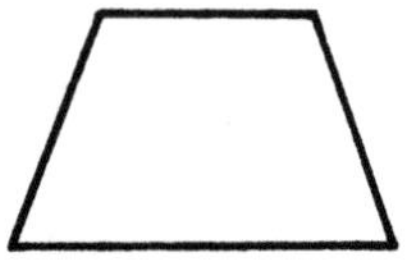

8.

9.

10.

11. Draw three similar triangles.

12. Draw three congruent pentagons.

Name ______________________

Describe Each Figure

A point where sides meet is called a vertex. Vertices means more than one vertex. Look at each figure. Count the sides. Count the vertices.

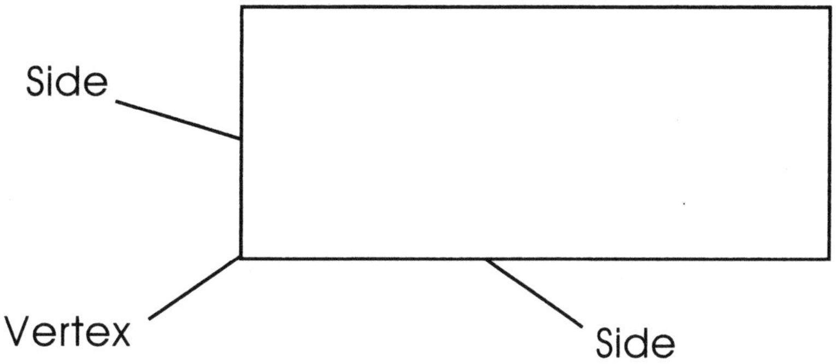

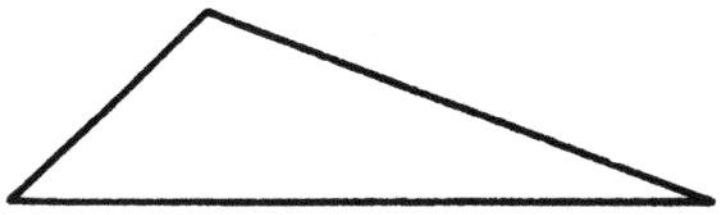

1. sides: ______________________
   vertices: ____________________

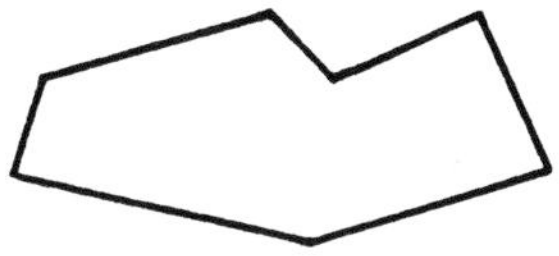

2. sides: ______________________
   vertices: ____________________

3. sides: ______________________
   vertices: ____________________

4. sides: ______________________
   vertices: ____________________

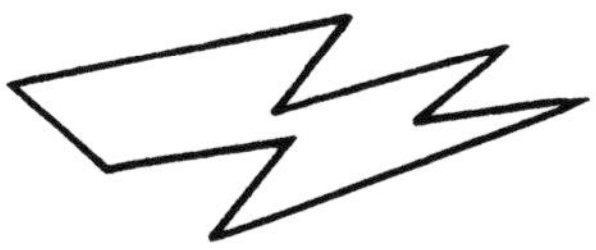

5. sides: ______________________
   vertices: ____________________

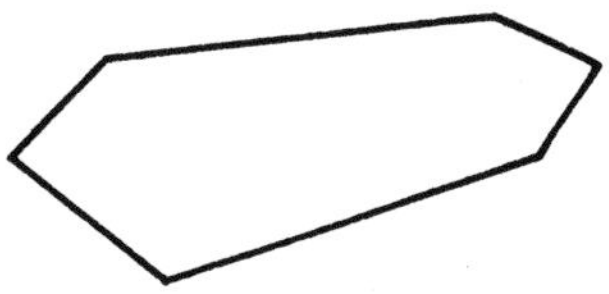

6. sides: ______________________
   vertices: ____________________

Name ______________________

Match Up

Match a word in the first column to a description in the second column to a picture in the third column.

hexagon

octagon

parallelogram

pentagon

polygon

rectangle

square

trapezoid

triangle

1. closed figure that has three or more sides
2. polygon that has 4 sides and 4 vertices
3. polygon that has 8 sides and 8 vertices
4. polygon that has 3 sides and 3 vertices
5. quadrilateral that has one pair of parallel sides
6. polygon that has 5 sides and 5 vertices
7. quadrilateral with opposite sides that are parallel
8. polygon that has 6 sides and 6 vertices
9. rectangle with 4 equal sides

Name ______________________

Quadrilaterals

A quadrilateral is any figure with 4 sides and 4 vertices. Some quadrilaterals have special names.

Trapezoid—a quadrilateral with one set of parallel sides
Parallelogram—a quadrilateral with two sets of parallel sides
Rectangle—a special parallelogram with four right angles and opposite sides of equal length
Square—a special rectangle with four equal sides

Put an X on any shape that is not a quadrilateral.
Circle all parallelograms.
Color all rectangles blue.
Color all trapezoids green.
Color all squares red.

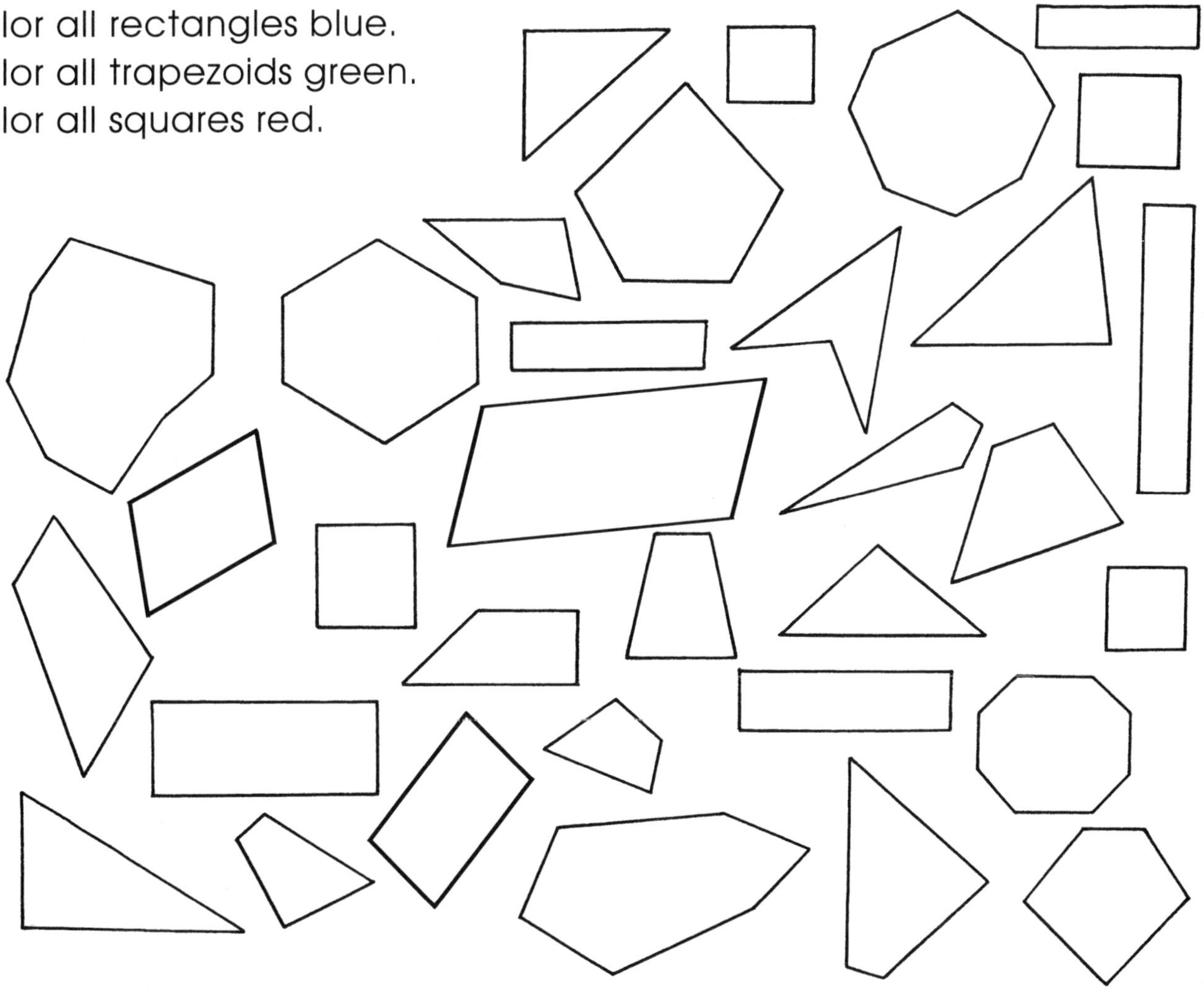

Name ______________

What's Your Angle

Use the definitions to label each angle below.

right angle: forms a square corner. It has a measure of 90°.
acute angle: has a measure that is less than a right angle (less than 90°)
obtuse angle: has a measure that is greater than a right angle (greater than 90°)

1. ______________

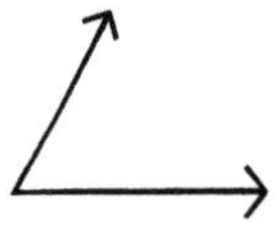

2. ______________

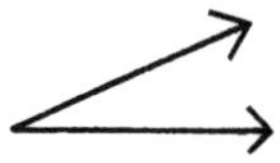

3. ______________

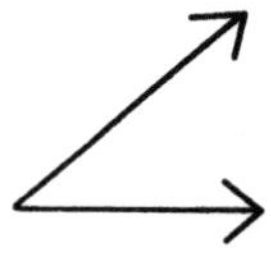

4. ______________

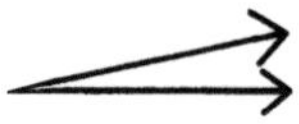

5. ______________

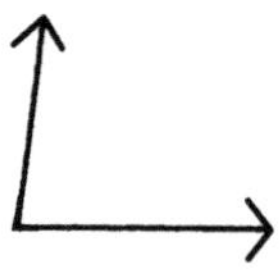

6. ______________

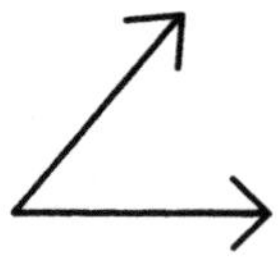

7. ______________

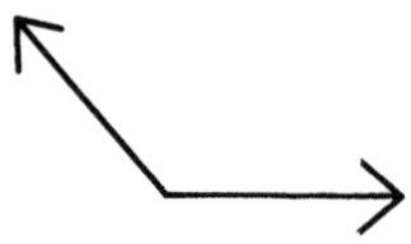

8. ______________

9. ______________

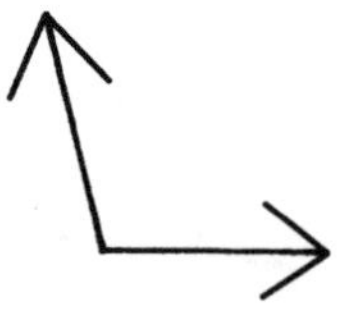

10. ______________

11. ______________

12. ______________

13. ______________

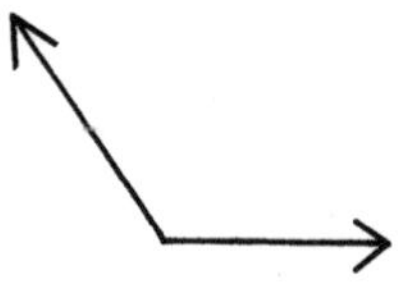

14. ______________

Name ____________________

Solids

Study the solids. Use what you know to fill in the chart. Remember, the flat surface of a solid figure is a face. A line segment where two faces meet is an edge. A point where edges meet is a corner.

Solid	Number of faces	Number of edges	Number of corners
cone			
rectangular prism			
cylinder			
octagonal prism			
cube			
triangular prism			
hemisphere			
sphere			
square pyramid			

Use the table above to help you name each mystery solid.

1. I have 0 faces or corners. You can find me at sporting events. What am I? ____________________
2. I have more than 3 faces. I have a total of 6 corners. My faces come in 2 shapes. I make great rainbows. What am I? __________
3. I have only 1 face. I have no corners, but I do have a point. I make summer delicious. What am I? ____________________
4. I have more than 2 edges and more than 2 corners. Each of my faces is exactly the same. Many of me stack easily. What am I?

   ____
5. My 2 faces have 0 edges. I have 0 corners. I make moving objects from one place to another easy. What am I? ____________________

 Name ____________

Dimensions

Match each three-dimensional shape with a corresponding two-dimensional shape by drawing a line. You may need to match two two-dimensional shapes to one three-dimensional shape.

1. 

2. 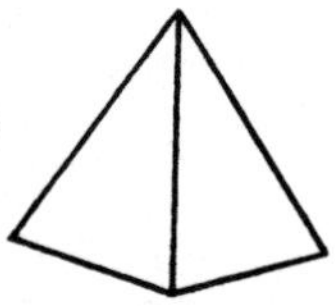

3. 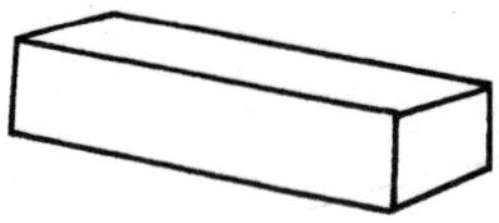

4. 

triangle

square

circle

rectangle

Look at each solid. Draw the two-dimensional shapes needed to make each shape. Shapes may be made of more than one two-dimensional shape.

5. 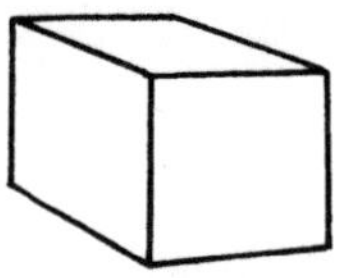

6. 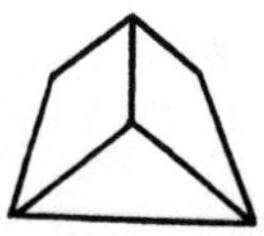

7. 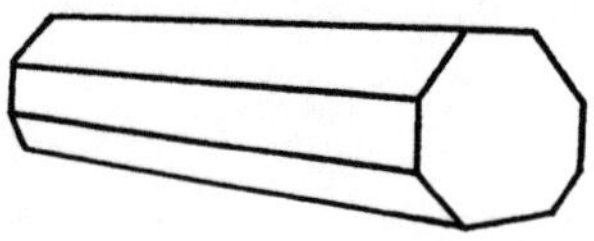

8. What two-dimensional object is representative of a sphere?

____

Name ______________________

Around the Rim

The distance around a figure is the perimeter. Find the perimeter of the following figures. Don't forget to label.

1.

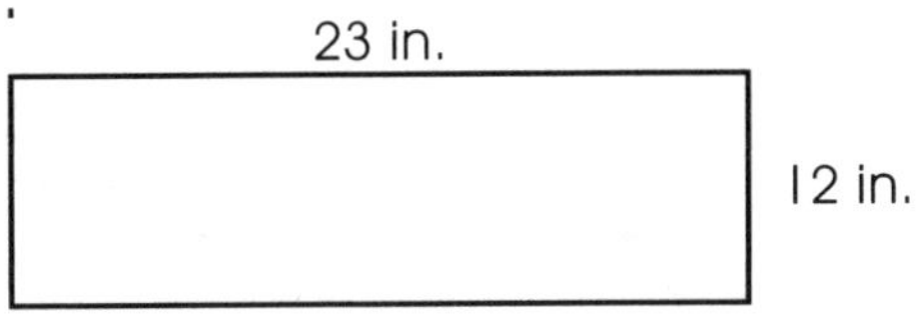

____

2.

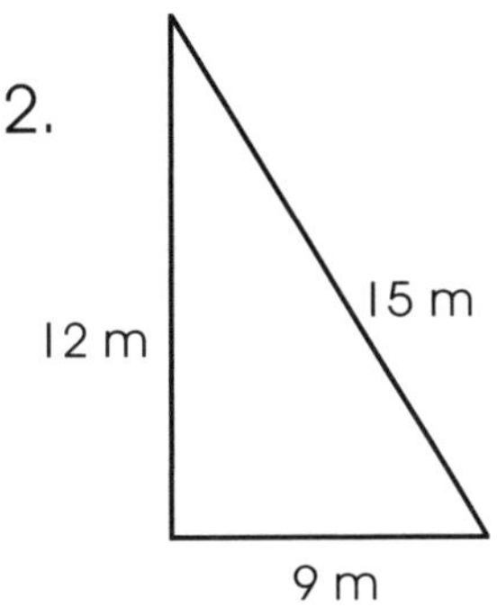

____

3.

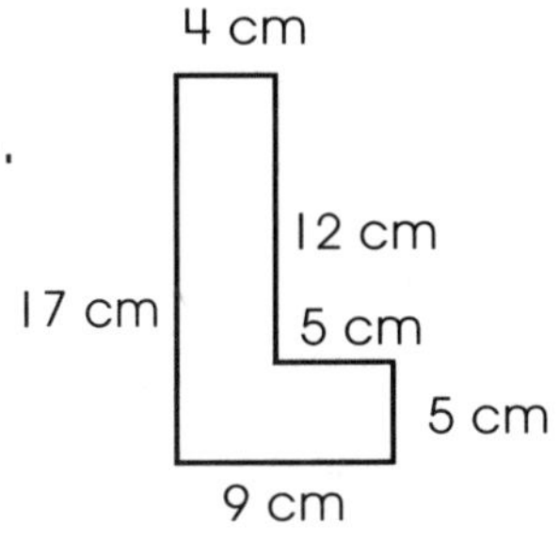

____

4.

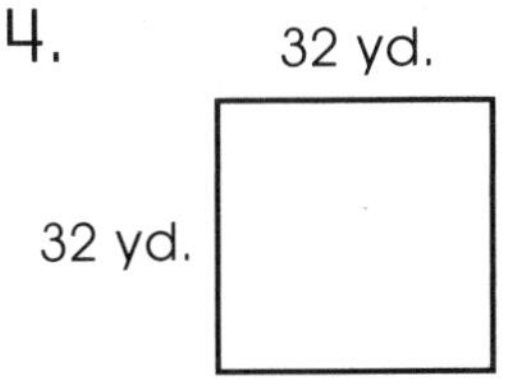

____

5.

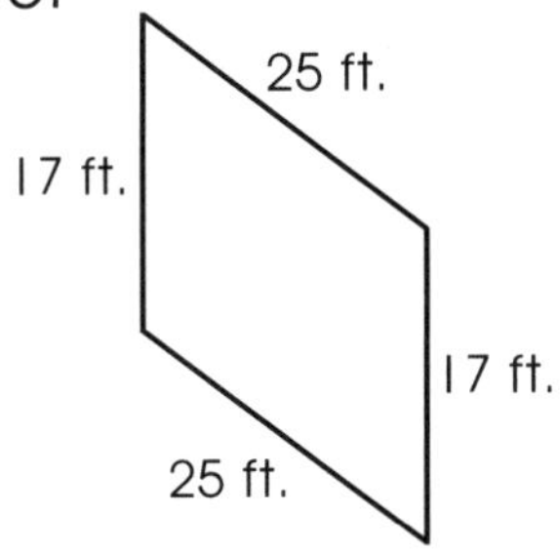

____

6.

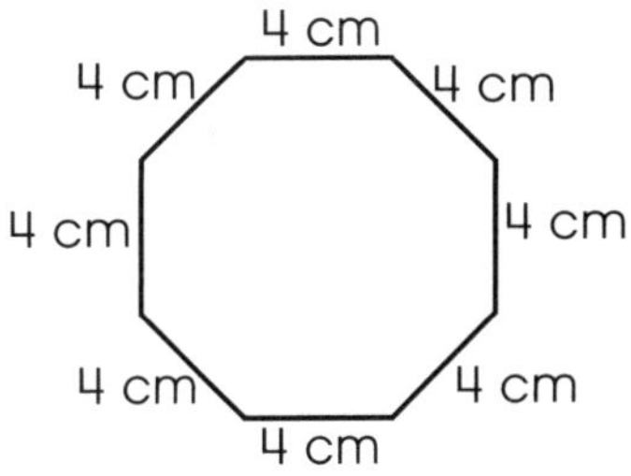

____

7.

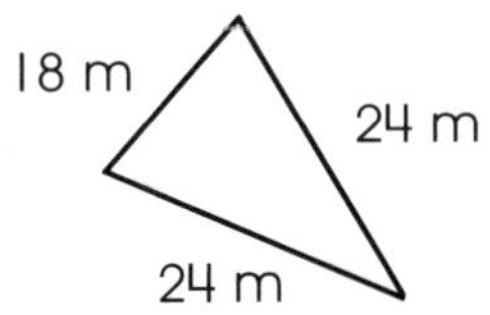

____

8.

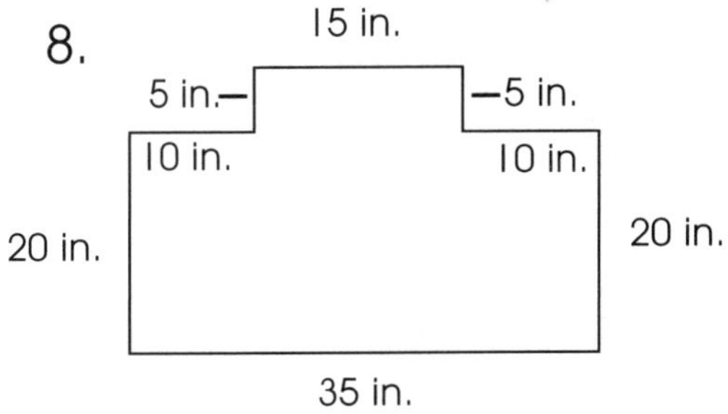

____

Name ____________________

Tiling

Kayla is working with tiles. Help find the area and perimeter of each figure. The number of square units that cover a shape is the area. It is recorded in square units. The distance around a shape is the perimeter. It is recorded in units.

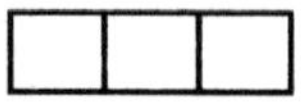

1.
perimeter = ____________
area = ________________

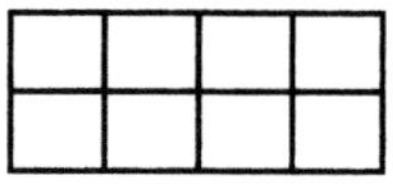

2.
perimeter = ____________
area = ________________

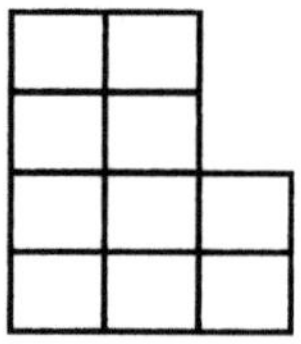

3.
perimeter = __________
area = ________________

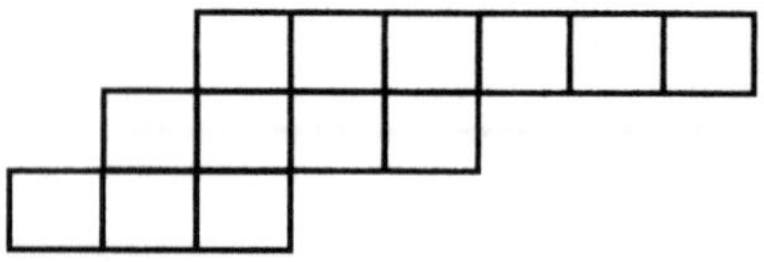

4.
perimeter = __________
area = ________________

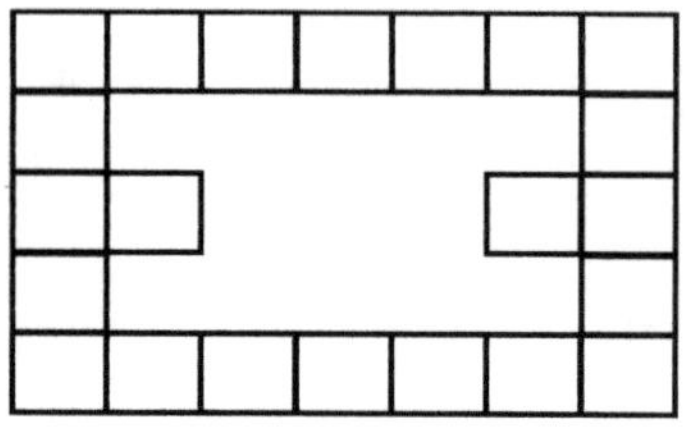

5.
perimeter = ____________
area = ________________

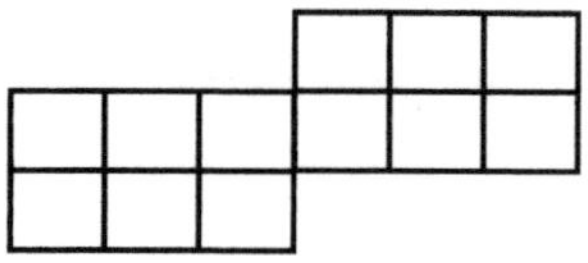

6.
perimeter = ____________
area = ________________

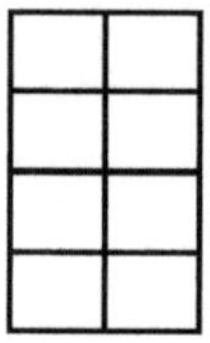

7.
perimeter = ____________
area = ________________

Name ____________________

Calculating Area

The rule for finding the area of a rectangle is **length x width**. Look at each figure drawn below. Use the length and width given to find the area. Label your answer in square units. Then, find the perimeter of each figure. Label with units.

1. ____________________
____

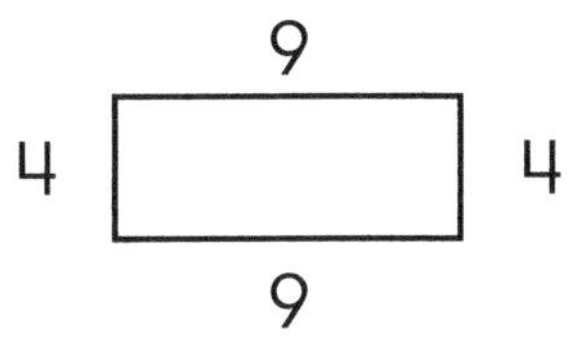

2. ____________________
____

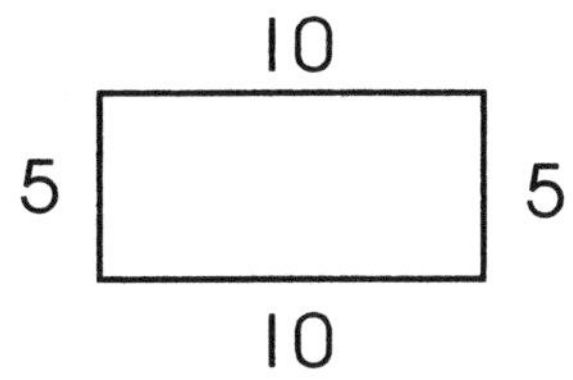

3. ____________________
____

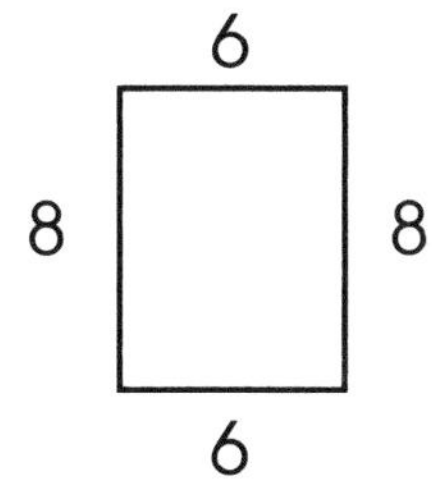

4. ____________________
____

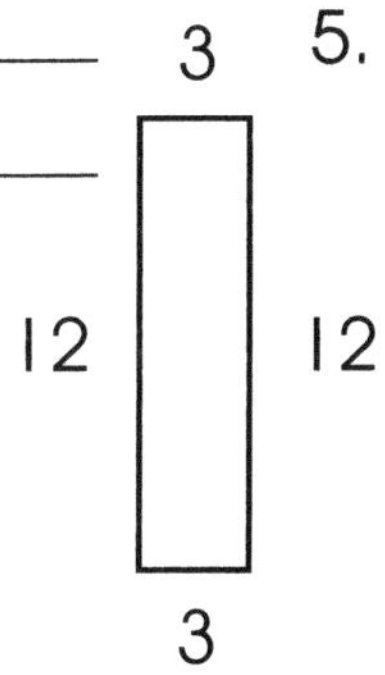

5. ____________________
____

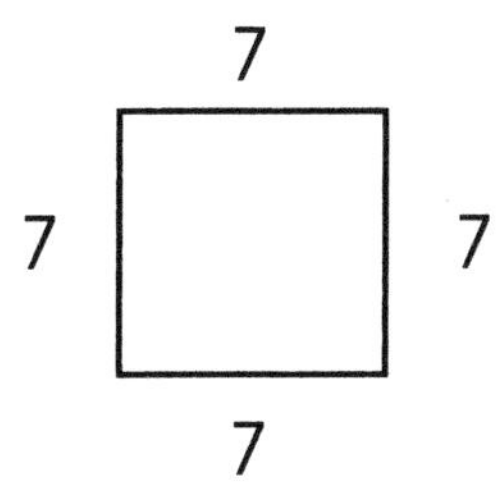

6. ____________________
____

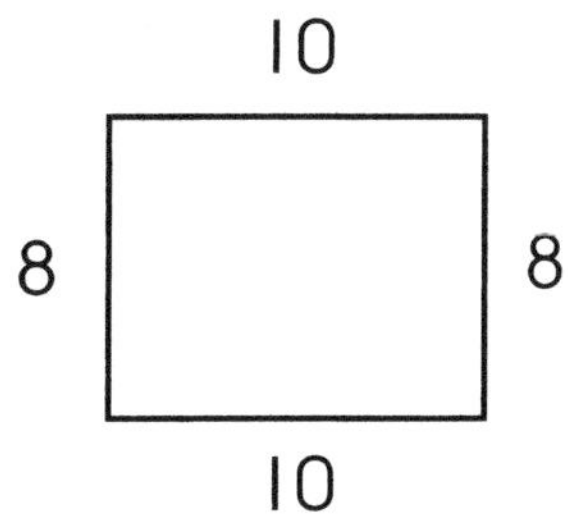

Solve each story problem.

7. Joey is covering the bottom of a wooden box with felt. The bottom of the box is 8 inches by 4 inches. How many square inches of felt does he need? ____________________

8. Tony wants to cover a trading card with contact paper. The card is 3 inches by 5 inches. How many square inches of contact paper does he need to exactly cover both sides?

____

Measurement—area of rectangle   Name ____________________

Cover It

The rule for finding the area of a rectangle is **length x width = area**. Look at each figure drawn below. Use the length and width given to find the area. Write your answer in square units.

1. ____________________

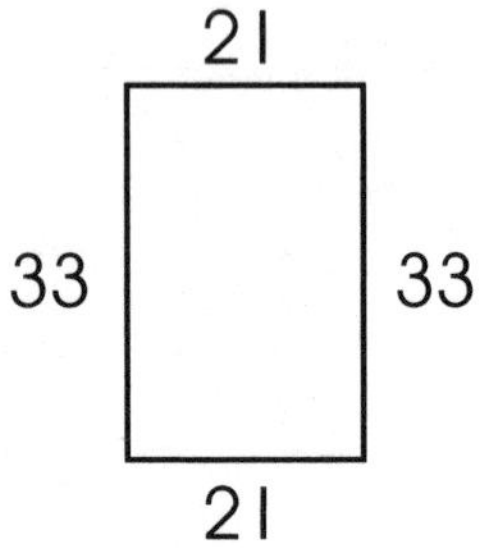

2. ____________________

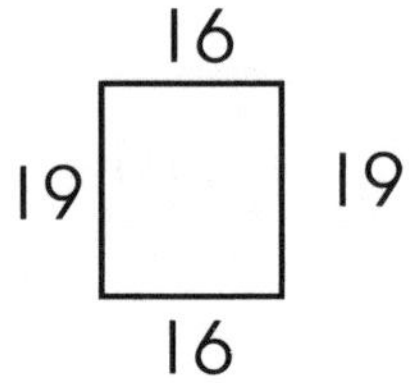

3. ____________________

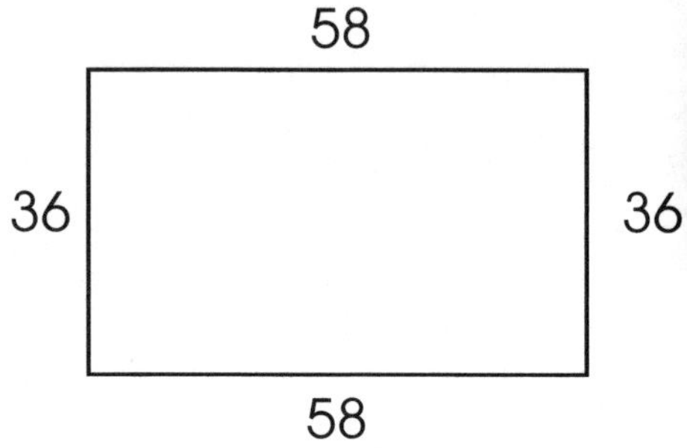

4. ____________________

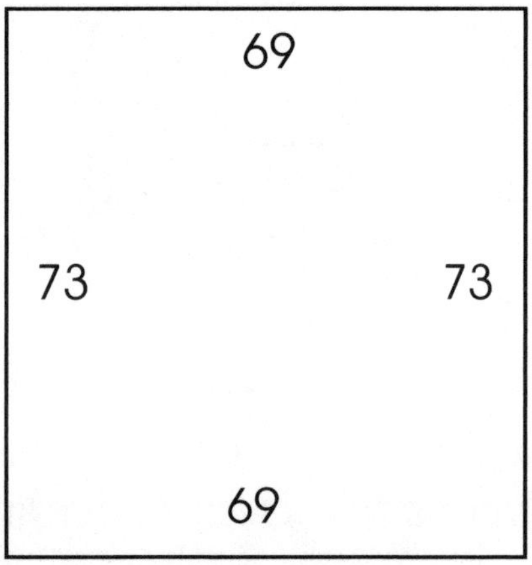

5. ____________________

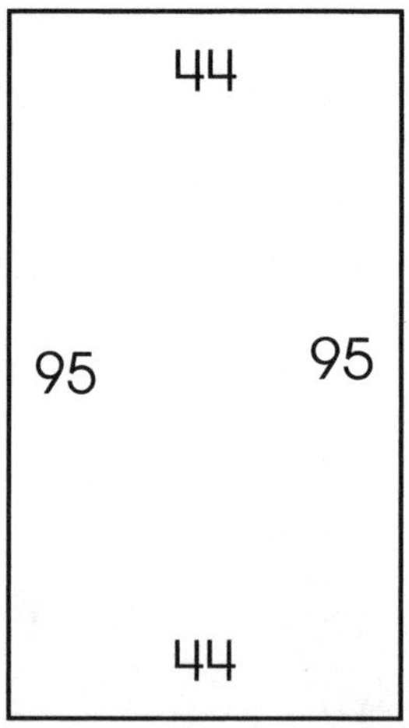

6. ____________________

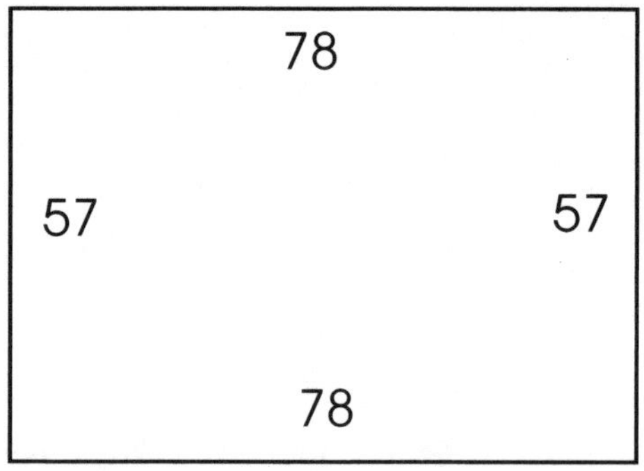

Write the answer to each problem in the correct part of the Venn diagram. Remember, there is a set outside the two circles.

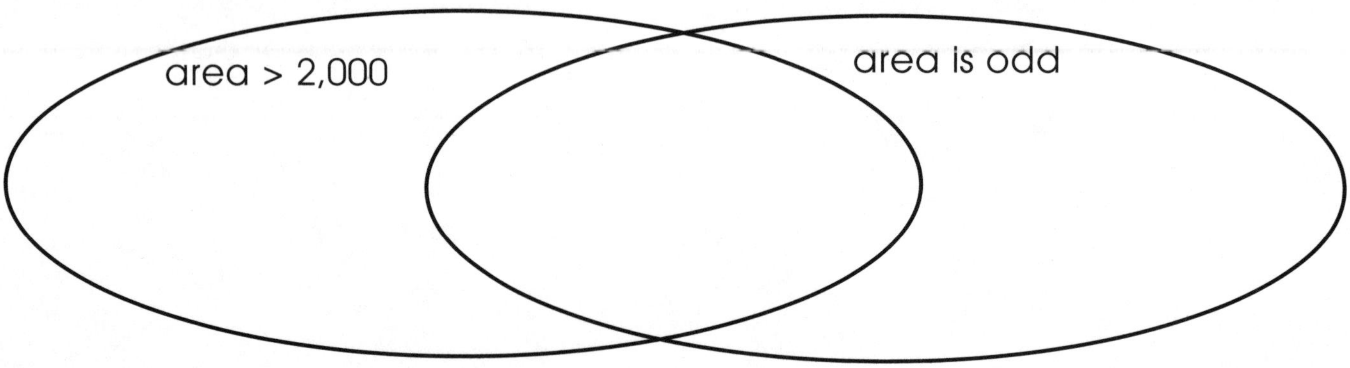

Name ____________________

Shadows

Determine the area of each figure.

Area of triangle: $\frac{1}{2}$ **base x height**

Area of rectangle: **length x width**

Area of parallelogram: **base x height**

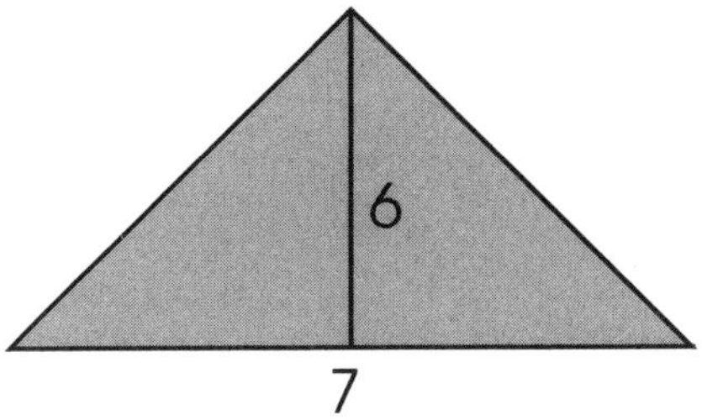

1. ____________________

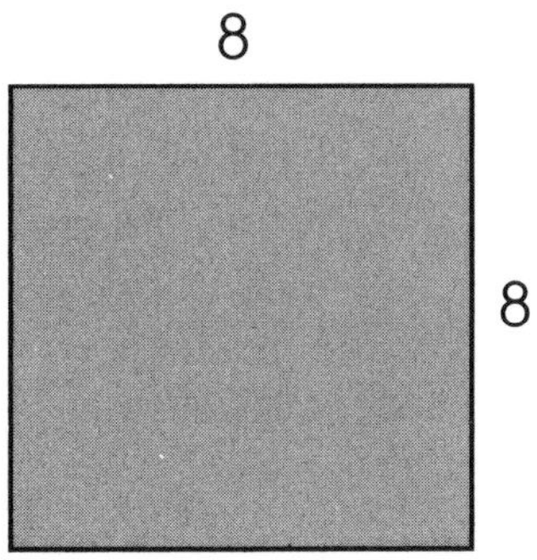

2. ____________________

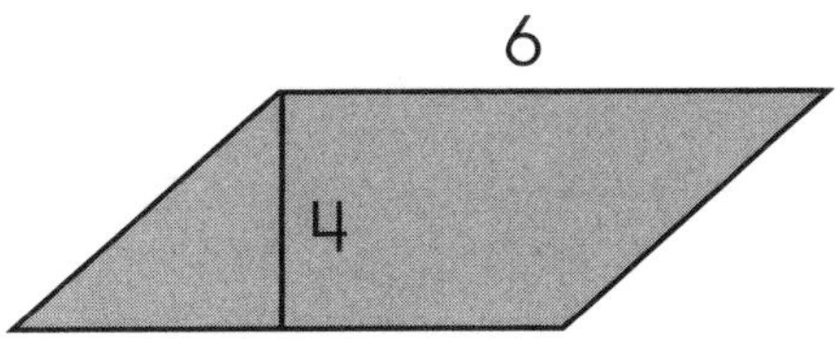

3. ____________________

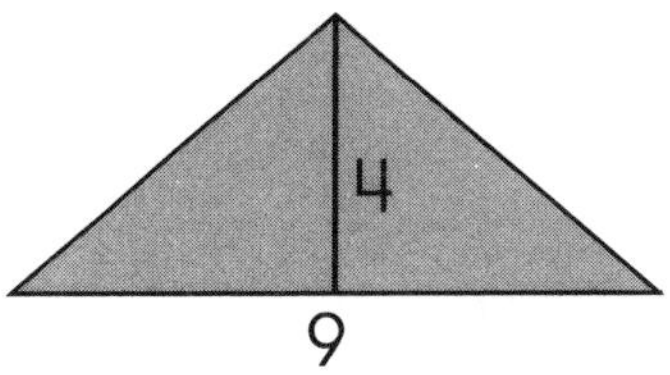

4. ____________________

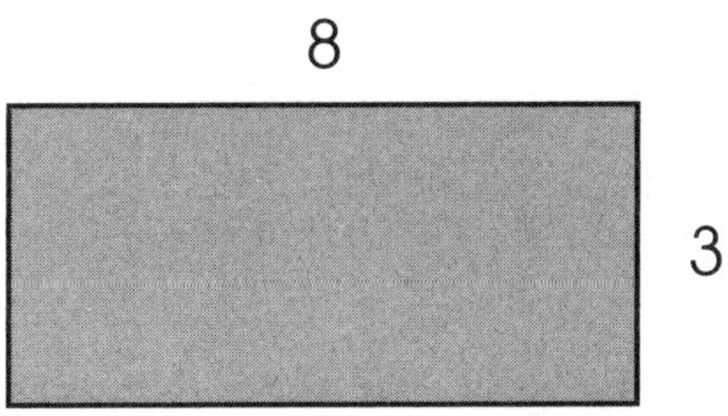

5. ____________________

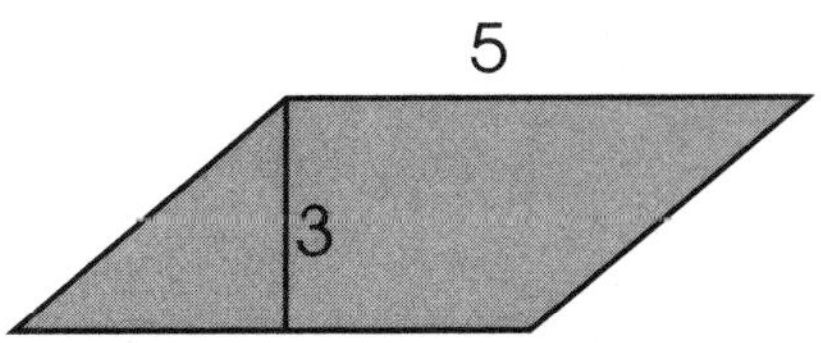

6. ____________________

Measurement—volume

Name ____________________

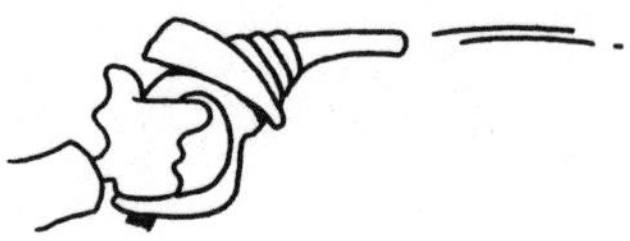

Fill It Up

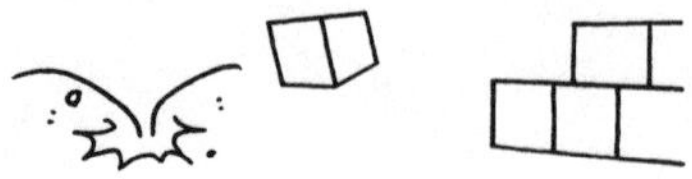

The number of cubic units of space inside a solid figure is its volume. Each block has a volume of 1 cubic unit. Fill in the length, width, and height for each figure. Find the volume of each figure. Record the volume in cubic units.

Volume = l x w x h

1. 

1. length = ____________
width = ____________
height = ____________
volume = ____________

2. 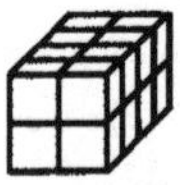

2. length = ____________
width = ____________
height = ____________
volume = ____________

3. 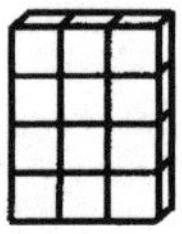

3. length = ____________
width = ____________
height = ____________
volume = ____________

4. 

4. length = ____________
width = ____________
height = ____________
volume = ____________

5. 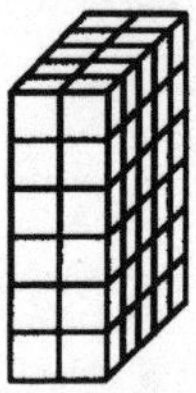

5. length = ____________
width = ____________
height = ____________
volume = ____________

6. 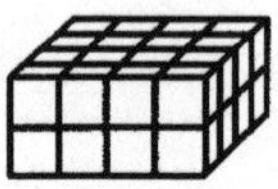

6. length = ____________
width = ____________
height = ____________
volume = ____________

Name ____________________

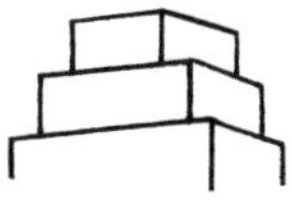

Upping the Volume

Find the volume of each container. Write your answer in cubic units.
Volume = l x w x h

1. ____________________

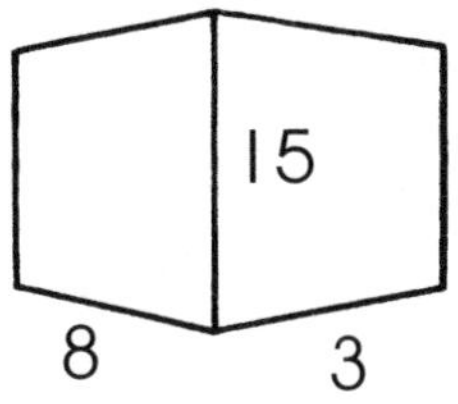

2. ____________________

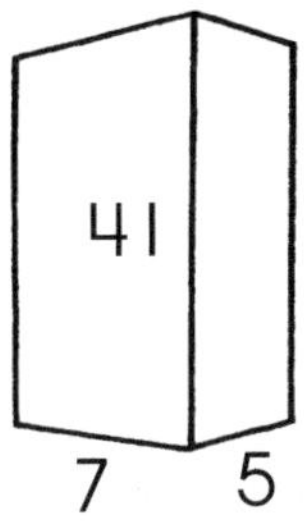

3. ____________________

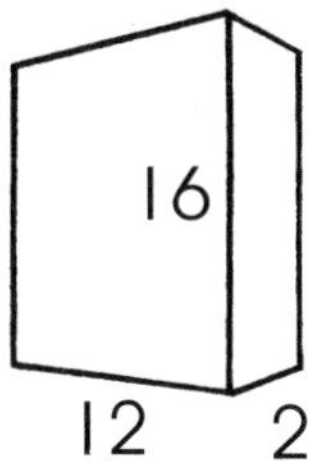

4. ____________________

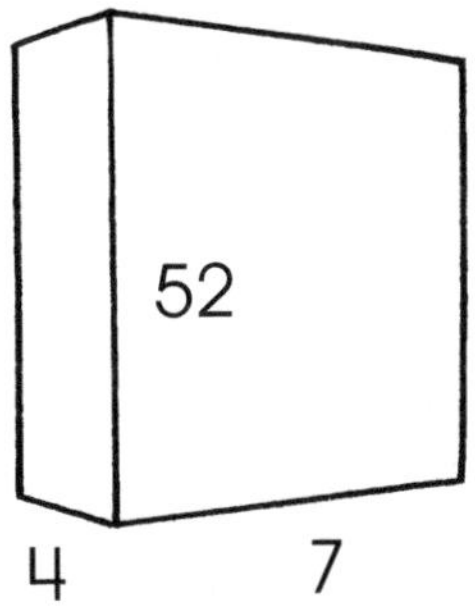

5. ____________________

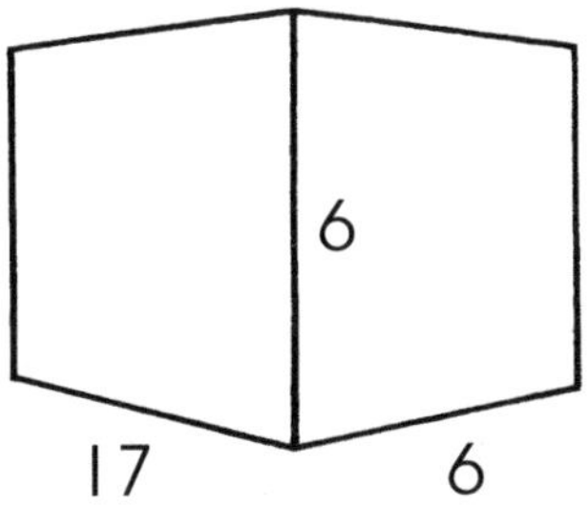

6. ____________________

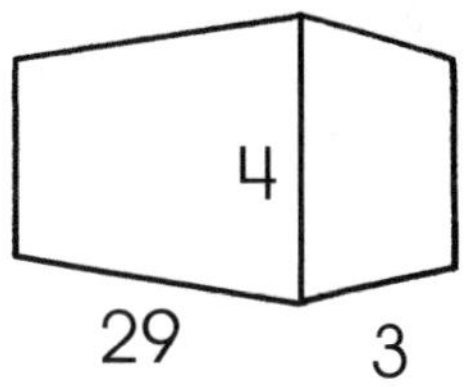

Read and solve each story problem. Show your work and label.

7. Willie's family wants to pour a cement slab for basketball. It will be 24 feet by 18 feet by 1 foot deep. How many cubic feet of cement will they need? ____________________

8. Maddie has 3 pans. The first is 8 by 8 by 2 inches. The second pan is 9 by 9 by 2 inches. The third pan is 9 by 14 by 2 inches. She needs to make 150 cubic inches of fudge. Which pan should she use? ____________________

Name ____________________

Wall of Snow

Five friends used different size containers to make snow bricks for building. Determine the volume of each container. Use the clues and the matrix to match the person with the container.

- Jen does not have the container with the least volume.
- The volume of Juri's container is greater than 450 cubic inches.
- Josh's container is 1 square foot.
- Jill's container has exactly half the volume of Josh's container.

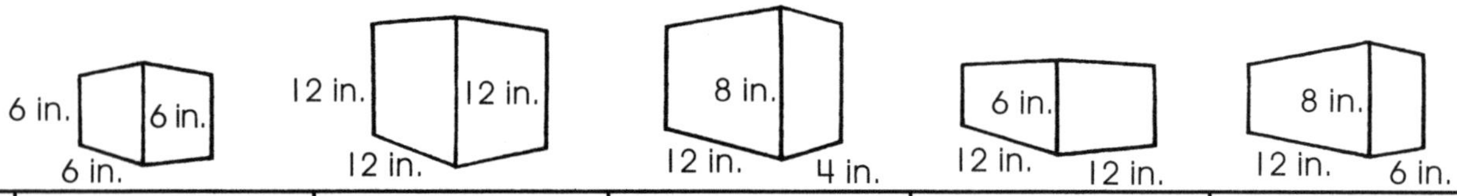

Jake					
Jen					
Jill					
Josh					
Juri					

Find the total volume of snow for each. Each wall is 1 block thick.

1. Jake built a wall that was 12 blocks across and 8 blocks high. ____________________

2. Jen built a wall that was 11 blocks across and 9 blocks high. ____________________

3. Jill built a wall that was 8 blocks across and 4 blocks high. ____________________

4. Josh built a wall that was 5 blocks across and 3 blocks high. ____________________

Name ____________________

From Here to There

Fill in the blanks in the measurement chart. Use it to answer the questions below.

English System

1 foot (ft.)	=	12 inches (in.)
1 yard (yd.)	=	_____ inches (in.)
1 yard (yd.)	=	3 feet (ft.)
1 mile (mi.)	=	_______ feet (ft.)
1 mile (mi.)	=	1,760 yards (yd.)

1. Jofie has an eagle kite. It is 52 inches long. How many feet and inches is this?

   ____

2. Allie walked 3 miles. How many yards did she walk? How many feet did she walk?

   ____

3. Sydney is running a mile for gym. She has gone 942 yards. How many yards does she still have to go?

   ____

4. A porpoise can measure 7 feet. How many inches is this?

   ____

5. A dolphin can measure 13 feet. How many yards and feet is this?

   ____

6. Ian has a $1\frac{1}{2}$-foot-wide space on his wall for a poster. He found a poster he likes that is 20 inches wide and 30 inches tall. Will the poster fit on his wall?

   ____

7. A humpback whale can measure 39 feet long. How many yards is this?

   ____

8. Emma crocheted a blanket that is 2 yards long. How many inches long is it?

   ____

9. José needs 8 yards of board. The boards come in 4-foot lengths. How many 4-foot boards does he need?

   ____

Name ______________

Measure This

Follow each set of directions to measure each line.

Measure each line to the nearest centimeter.

1. ______________________
2. ________________
3. ____
4. ______
5. ____

Measure each line to the nearest quarter inch.

6. ______
7. ____
8. ______________________________
9. ___________________
10. ____

11. Draw a line 10 centimeters long. Label it.

12. Write the name of 1 object in the room shorter than your 10-centimeter line.

    ____

13. Write the name of 1 object in the room longer than your 10-centimeter line.

    ____

14. Draw a line $5\frac{2}{4}$ inches long. Label it.

    ____

15. Write the name of 1 object in the room shorter than your $5\frac{2}{4}$ -inch line

    ____

16. Write the name of 1 object in the room longer than your $5\frac{2}{4}$ -inch line.

    ____

 Name ____________________

Metric Length

Fill in the blanks in the measurement chart. Use it to answer the questions below.

Metric System

1 centimeter = 10 millimeters	1 decameter = _____ centimeters
1 decimeter = 10 centimeters	1 hectometer = _____ decameters
1 meter = 10 decimeters	1 hectometer = 100 meters
1 meter = _____ centimeters	1 kilometer = _____ hectometers
1 meter = _____ millimeters	1 kilometer = 100 decameters
1 decameter = 10 meters	1 kilometer = 1000 meters
1 decameter = _____ decimeters	

1. Alfie wants to measure the length of the playground. Which unit of measure should he use and why?

   ____
   ____
   ____

2. Dan needs to measure the length of his arm. Which unit of measure should he use?

   ____

3. Jordi measured her tomato plant. It is 34.5 centimeters. How many millimeters is this? How many decimeters?

   ____
   ____

4. Anna is walking in a 5-kilometer charity event. She has just passed the 342-decameter mark. How many decameters away is the finish line? How many hectometers?

   ____
   ____

5. Meg has a plastic case that is 4 centimeters by 4 centimeters. She found a shell that is 34 millimeters long. Will it fit in her case? ______________

6. Kifa jumped 1.27 meters. How many centimeters is this? How many decimeters is it? How many millimeters is it? __________

   ____

Measurement—mass

Name ____________________

Tonnage

English System	Metric System
1 pound (lb.) = 16 ounces (oz.)	1 gram (g) = 1,000 milligrams (mg)
1 ton (T) = 2,000 pounds (lb.)	1 kilogram (kg) = 1,000 grams (g)
	1 metric ton (t) = 1,000 kilograms (kg)

Complete the word problems using the measurement tables above.

1. Gina has a 5-pound bag of cement. How many ounces is this? ____________________

2. Rick needs 3 pounds of chocolate chips. They come in 12-ounce bags. How many bags must he buy?

   ____

3. A humpback whale can weigh 39 tons. How many pounds is this? ____________________

4. White rhinos can weigh 5,500 pounds. What is the smallest possible ton-weight truck that could deliver the white rhino to the zoo? ____________________

5. A large sculpture was donated to the park. It weighs 3 metric tons. The base is 2,468 kilograms. How much does the rest weigh?

   ____

6. Tallie has to mix feed for her horse. Feed comes in 2-kilogram containers. She needs 786 grams of wheat and 648 grams of oats. She just opened new containers. How much wheat and oats are left in the original containers after she feeds the horses?

   ____

7. Jen takes a 1-gram vitamin. It has 60 milligrams of vitamin C, 100 milligrams of calcium, 18 milligrams of iron, 15 milligrams of zinc, and 2 milligrams copper. How many milligrams are these 5 elements? How much is left over for other items in the vitamin?

   ____

Name ______________________

Measuring Mass

Look at the table. Use it to answer the following questions. Show your work.

Object	Estimated Mass	Actual Mass
cup	16 grams	25 grams
washer	8 grams	6 grams
three balloons	18 grams	15 grams
two clips and washer	9 grams	14 grams
cup and water	40 grams	86 grams
balloon and air	8 grams	6 grams

1. What is the actual mass of 1 washer? Compare the actual to the estimated mass.

   ____

2. Compare the estimated to the actual mass of the cup and water. ______________

3. What is the actual mass of 3 balloons? Use this information to find the mass of 1 of the balloons.

   ____

4. What is the difference between the estimated and actual mass of the cup?

   ____

5. What is the actual mass of 1 clip?

   ____

6. What is the actual mass of the air in the balloon? __________

7. What is the actual mass of the water? ______________

8. If all of the water in the cup was poured into 1 balloon, what would be the actual mass of the balloon filled with the water? ____________

9. What would be the actual mass of 7 clips? ____________

Name ____________________

Capacity

Fill in the missing amounts in the chart. Use it to answer the questions.

1 tablespoon = 3 teaspoons	1 quart = 2 pints
1 cup = 16 tablespoons	1 quart = ____ fluid ounces
1 cup = ____ teaspoons	1 gallon = ____ cups
1 cup = 8 fluid ounces	1 gallon = ____ pints
1 pint = 2 cups	1 gallon = 4 quarts
1 pint = ____ fluid ounces	1 gallon = ____ fluid ounces
1 quart = ____ cups	

1. Mikaela has 5 people in her family. Each person drinks six 8-fluid ounce glasses of water a day. They keep cold water in the refrigerator. Is a 1-gallon jug of water enough for the day?

   ____

2. George and Hale are at the school carnival. Soda costs 30¢ per cup with your own container. George has a 16-ounce water bottle. How much does it cost to fill his bottle with pop? ________ Hale has a 32-ounce water bottle. How much does it cost to fill Hale's bottle? __________

3. Paula has a 20-gallon fish tank. She needs to treat the tank with a chemical. The directions say to put 1 drop in for every quart of water. How many drops of chemical are needed for this aquarium?

   ____

4. Lionel is baking. His recipe calls for 3 cups of sugar. The measuring cup is dirty, but not the measuring spoons. How many tablespoons of sugar should he use for this recipe? __________ What if he used teaspoons instead? __________

5. Jackie drinks 1 pint of chocolate milk each day with her school lunch. How many cups of chocolate milk is that in 1 week? ________ Does she drink enough in 1 week to fill a gallon? __________

 Name ____________________

Metric Capacity

Use the measurement chart to answer the questions below.

Metric System

1 teaspoon (tsp.) = 5 milliliters (mL)
1 tablespoon (tbsp.) = 15 milliliters (mL)
1 liter (L) = 1,000 milliliters (mL)
1 liter (L) = 1,000 cubic centimeters (cm^3)
1 liter (L) = 1 cubic decimeter (dm^3)
1 liter (L) = 4 metric cups
1 kiloliter (kL) = 1,000 liters (L)

1. Angel's recipe calls for 2 teaspoons of vanilla. How many milliliters of vanilla should he use?

   ____

2. Ronny needs 15 milliliters of medicine. How many teaspoons should she take?

   ____

3. There are 48 teaspoons in an English-system cup. Jessie's recipe calls for 2 English-system cups of milk. How many milliliters is this? Could he put in a liter of milk?

   ____

4. Emma has a planter that holds 2,360 cubic centimeters of soil. How many cubic millimeters is this? How many cubic decimeters is this?

   ____

5. Fred's pool holds 6,400 liters of water. How many kiloliters is this?

   ____

6. Elma has a 4-liter bottle of glue for refilling smaller bottles. The small bottles hold 200 milliliters. How many small bottles can she fill from the big one?

   ____

Measurement—temperature Name ____________________

Brrrr, It's Cold!

Read each thermometer. Write the temperature in Fahrenheit and Celsius.

1. ________ °F

________ °C

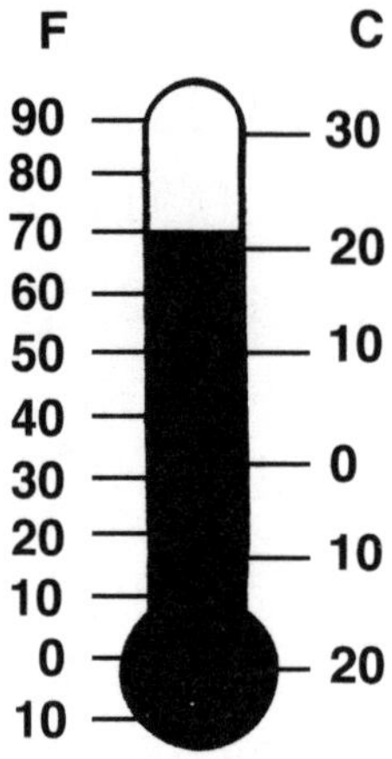

2. ________ °F

________ °C

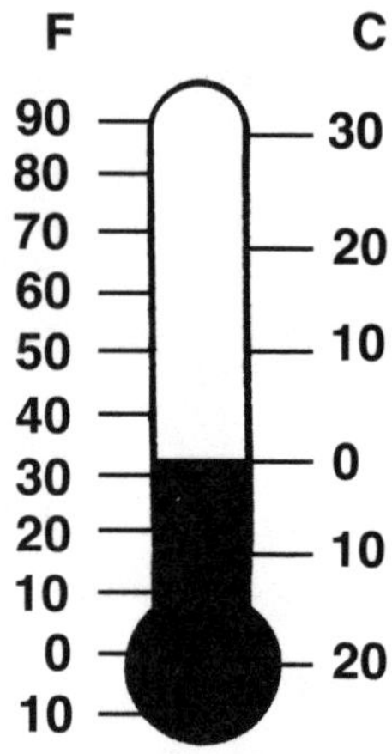

3. ________ °F

________ °C

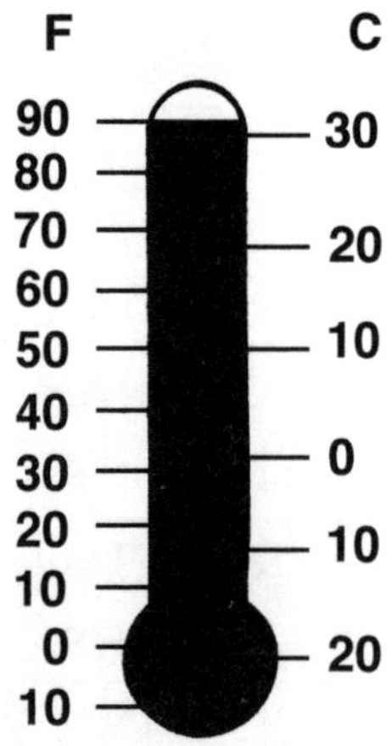

4. ________ °F

________ °C

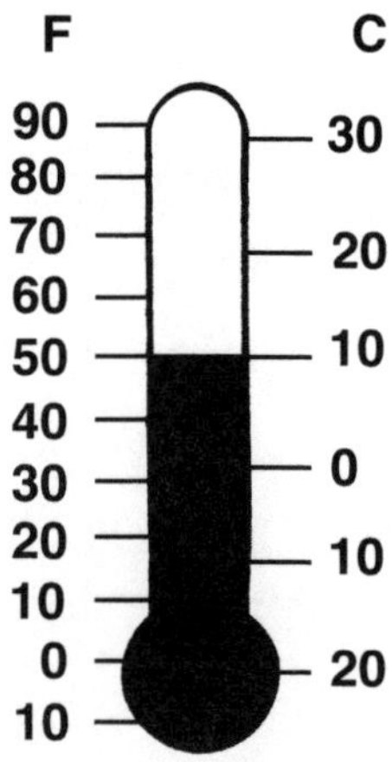

5. ________ °F

________ °C

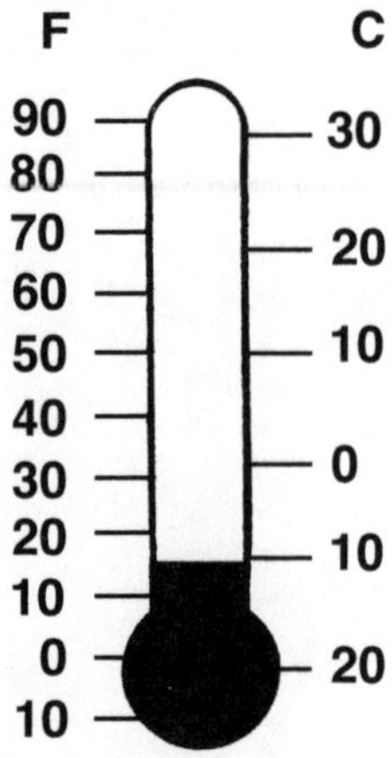

6. ________ °F

________ °C

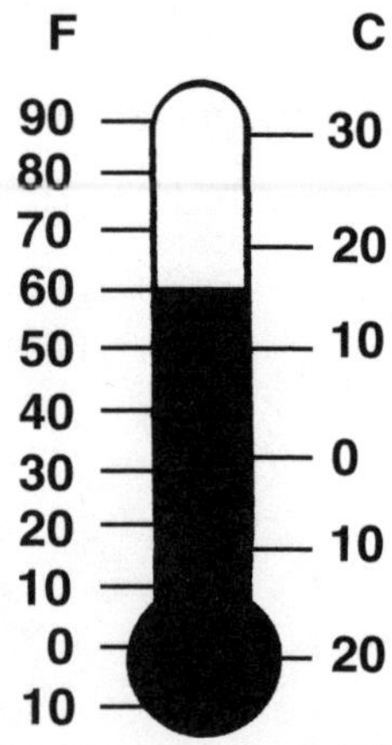

Name ____________________

What Time?

Write the time. Answer the questions.

1. 
What time is it? ____________
What time will it be in 1 hour? ______________

2. 
What time is it? ____________
What time was it 1 hour ago? ______________

3. 
What time is it? ____________
What time was it 20 minutes ago? __________

4. 
What time is it? ____________
What time will it be in 55 minutes? __________

5. 
What time is it? ____________
What time was it 45 minutes ago? __________

6. 
What time is it? ____________
What time will it be in 1 hour and 5 minutes? __________

7. 
What time is it? ____________
What time was it 2 hours ago? ______________

8. 
What time is it? ____________
What time was it 50 minutes ago? __________

Name ____________________

On Time

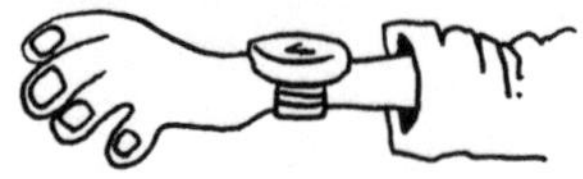

1. Nicole told Amy she would be at her house at 4:05. It takes her 15 minutes to get there on her bike. She left her house at 3:55. Will Nicole be early, on time, or late? By how many minutes?

   ____

2. Rowan needs to arrive at school at 8:10 for computer class. It takes him 20 minutes to walk. What is the latest time he could leave his house? ____________________

3. Meg has been playing her hand-held computer game since 3:55. She is allowed 40 minutes each day. It is 4:15. Can she play longer, or is it time to get off?

   ____

4. Johanna's softball game starts at 6:15. The game lasts 1 hour. Her family plans to go out for ice cream afterwards. Ice cream and the trip home take 35 minutes. What time will Johanna get home?

   ____

5. It is 7:10. José remembers he has homework to do before he can watch TV. He has a math activity that takes 15 minutes, and he has 20 minutes of reading. His favorite TV show starts at 7:30. Will he be able to start the program? If not, how many minutes will he miss?

   ____

6. Danzel wants to go swimming with his friends at the pool. They plan to pick him up at 1:15. Before leaving, he needs to finish his chores. He has to take out the garbage, which takes 7 minutes, empty the dishwasher, which takes 6 minutes, and clean his room, which takes 37 minutes. What time does he need to start if he wants to leave for the pool on time?

   ____

Name ____________________

Have the Time?

1. Willie's bus comes at 8:20. It takes him 20 minutes to brush his teeth and clean up. It takes 10 minutes to get dressed and put his school bag together. Eating takes 15 minutes. Getting his jacket and walking to the bus stop take 10 minutes. How long do the morning activities take? ________________ What is the latest time Willie must get up in order to get on the bus? __________

2. Mikaela has a hand-held computer trading card game. She meets someone at the park who is willing to make 10 swaps. She is getting ready to start when her battery gives a 4-minute warning. Any cards in the swap center when the power goes out will be permanently lost. It takes 15 seconds to get into the swap center and 20 seconds to get out of the swap center and save the trade. Each swap takes 25 seconds. How many seconds are there in 4 minutes? _________ How long does she have if she subtracts her in and out times? __________ How many 25 second trades can Mikaela safely make before her battery dies and all the cards are lost? _______________

3. The morning language arts block is from 8:45 to 10:30. Any incomplete work is done during choice time. Manuel's class went to gym from 9:10 to 9:45. It takes Manuel at least 8 minutes to do his spelling assignment and 12 minutes to edit yesterday's writing. He also has a 15-minute reading assignment and must work on a language arts computer program for 9 minutes. He meets with his reading group for 20 minutes. How long is the language arts block? _____________ How much time is left if the gym time is subtracted? _____________ How long do Manuel's morning activities take? _________ Will Manuel need to finish his work during choice time? ___________

 Name ____________

August

Maddie is looking over her August calendar. Help her answer the following questions.

Sunday	Monday	Tuesday	Wednesday	Thursday	Friday	Saturday
			1	2	3	4
5	6	7	8	9	10	11
12	13	14	15	16	17	18
19	20	21	22	23	24	25
26	27	28	29	30	31	

1. Maddie is paid for doing chores each Friday. How many times will she be paid in August? ____________
2. Maddie is going to the zoo on the fourth Thursday. What is the date? ____________
3. Maddie's birthday is August 21. What day of the week is that? ____________
4. Maddie's family is leaving for vacation on the first Tuesday. What is the date? ________
5. Her family is returning from vacation on August 15. How many days will they be gone? ____________
6. Maddie's dad is planning to take her to the dentist the Wednesday before they leave for vacation. What is the date? ____________
7. School starts on the last Tuesday. What is the date? ____________
8. Maddie is having a sleepover the Friday before school starts. What date is the sleepover? ____________
9. Maddie's school has off on August 31. What day of the week is that? ____________
10. How many days will Maddie be in school in August? ____________

Measurement—word problems Name ______________________

Hummingbirds

Solve these problems about hummingbirds.

1. A broadtail hummingbird's wings were clocked at 2,280 beats per minute. How many beats are in 5 minutes?

   ____

2. A male broadtail hummingbird will fly to a feeder 40 times an hour to fight off other hummingbirds. If its wings beat 2,280 times per 1-minute flight, how many times will its wings beat in an hour?

   ____

3. A male hummingbird makes a courtship power dive by flying 60 feet up into the air and diving straight down. In 1 hour it makes 45 courtship power dives. How many feet up is this in 1 hour?

   ____

4. A hummingbird's heart can beat more than 1,200 times per minute while flying. How many beats would that be in 4 minutes? ______________

5. A hummingbird must visit around 987 flowers each day to drink its weight in nectar. How many flowers would this be in a week?

   ____

6. A hummingbird needs from 7 to 12 kilocalories of energy a day. One kilocalorie is 1,000 calories. How many calories is 7 kilocalories? ______________
   If a hummingbird gets 6,289 calories, how many more are needed?

   ____

7. A hummingbird eats most of its calories at dusk. If it eats 123 calories each minute, how many calories is that in 1 hour? ______________________

8. A bird watcher sees 11 different hummingbird species each day. How many hummingbird species would that be in 2 weeks? __________

Name ____________________

Ready, Set, Go!

Madalen timed the number of seconds it took her to do a variety of activities. Use the graph to answer the questions.

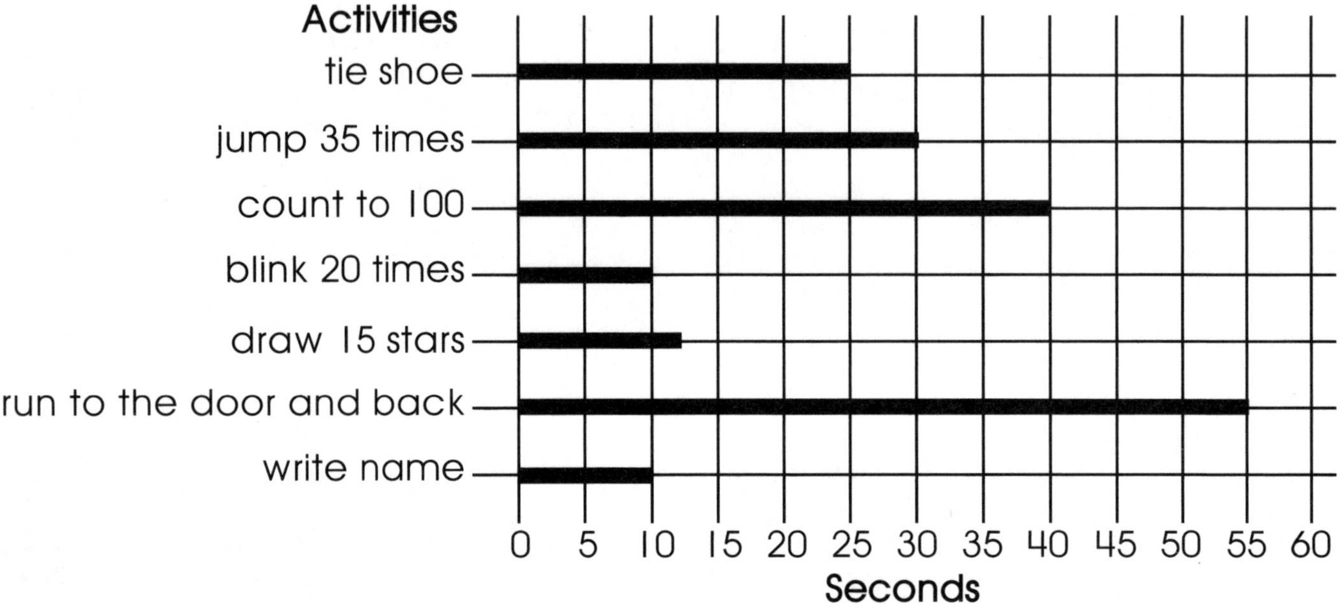

Circle the frequency table that matches the graph.

tie shoe	25
jump 35 times	30
count to 100	40
blink 20 times	10
draw 15 stars	12
run to the door and back	55
write name	10

count to 100	40
jump 35 times	30
draw 15 stars	10
run to the door and back	55
tie shoe	25
write name	10
blink 20 times	5

Answer the following questions assuming that Madalen will not get tired and slow down.

1. How many times could she write her name in 1 minute? ____________
2. How much faster could she draw 15 stars than blink 20 times? ______
3. Which 2 activities took an equivalent amount of time? ______________

   ____
4. In 2 minutes, how many times could she count to 100? ______________
5. What's the difference between the amount of time it took to blink 20 times and the amount of time it took to run to the door and back?

   ____

Data Analysis and Probability—organizing data/Venn diagram

Name ____________________

Round About

Solve. Place the differences in the correct parts of the Venn diagram. Place a star next to each problem whose answer goes in the outside set.

1. 342 − 181	2. 764 − 236	3. 822 − 571	4. 963 − 348	5. 536 − 175
6. 412 − 363	7. 703 − 416	8. 883 − 285	9. 654 − 139	10. 380 − 296
11. 900 − 234	12. 531 − 167	13. 786 − 479	14. 804 − 237	15. 238 − 142
16. 517 − 349				

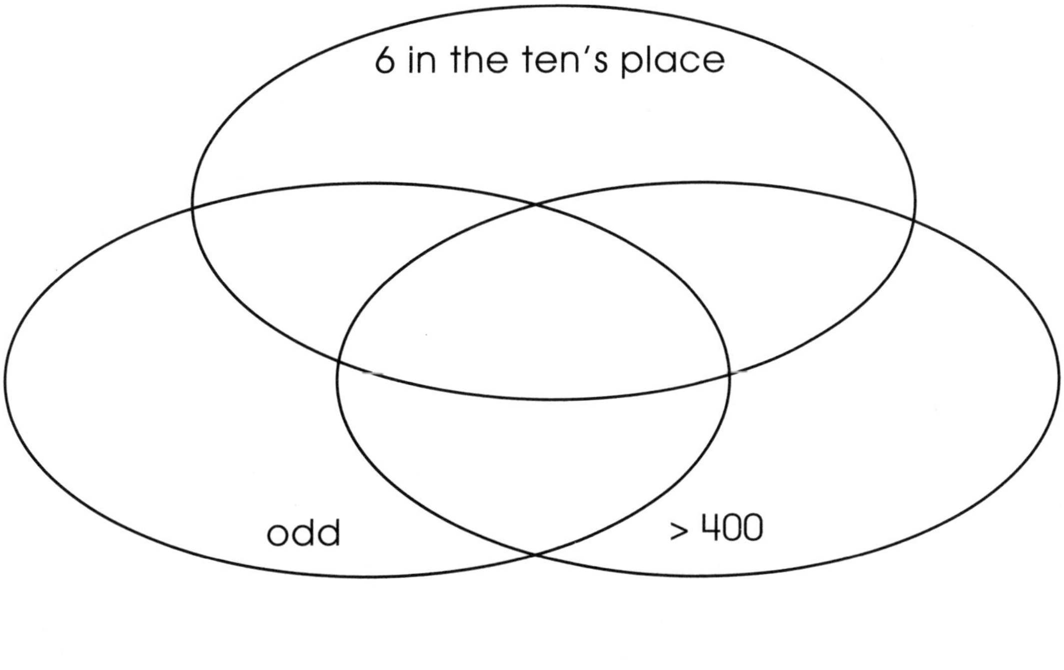

Name ______________________

Line Up!

1. Look at the line graph. What can you say about the pattern of the temperature throughout the day? Explain why you think this pattern occurs.

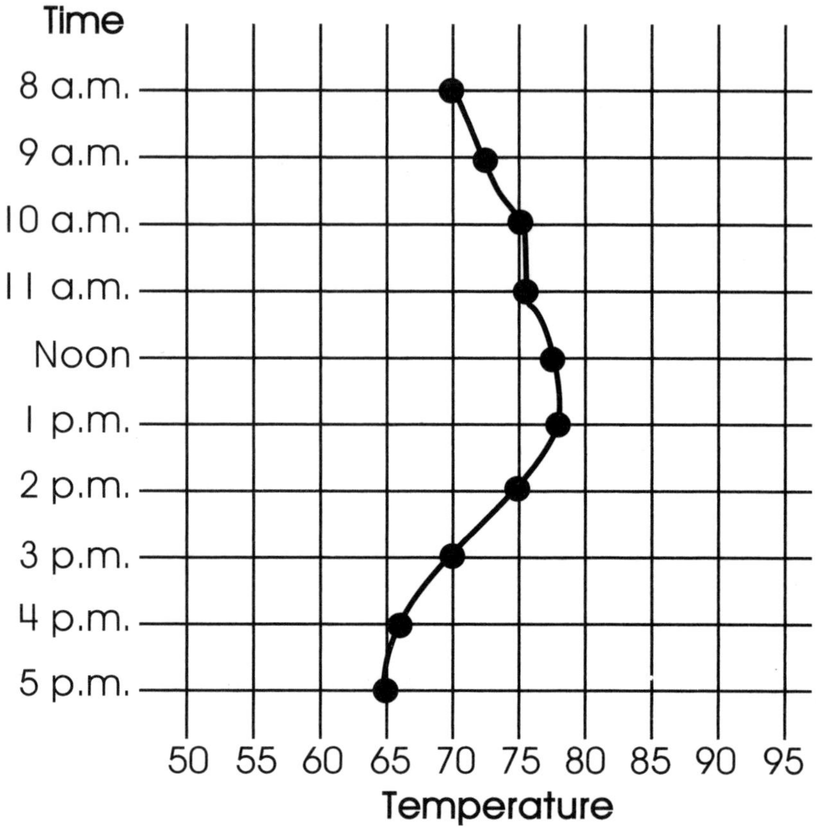

2. Look at the information. Plot the information in the graph below. Then divide the circle graph according to the percentages.

Gumdrops	3	10%	Tart Candies	6	20%
Candy Bars	18	60%	Lemon Drops	3	10%

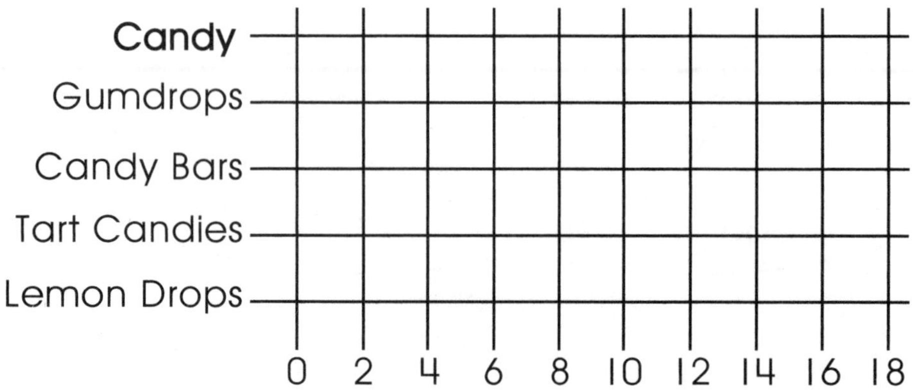

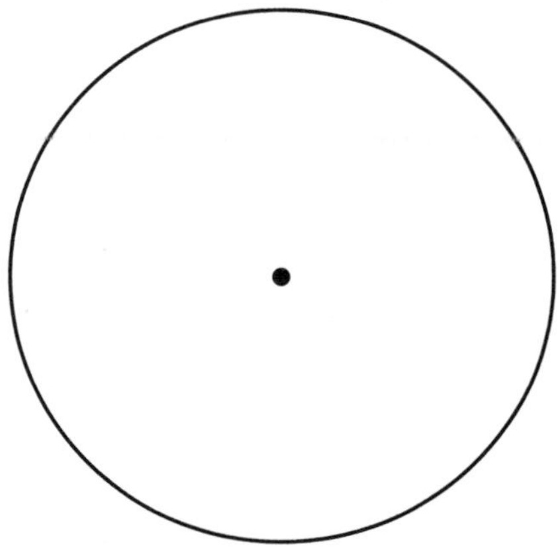

 Name ____________________

Softball

Use the softball tournament scoring matrix to answer the questions below.

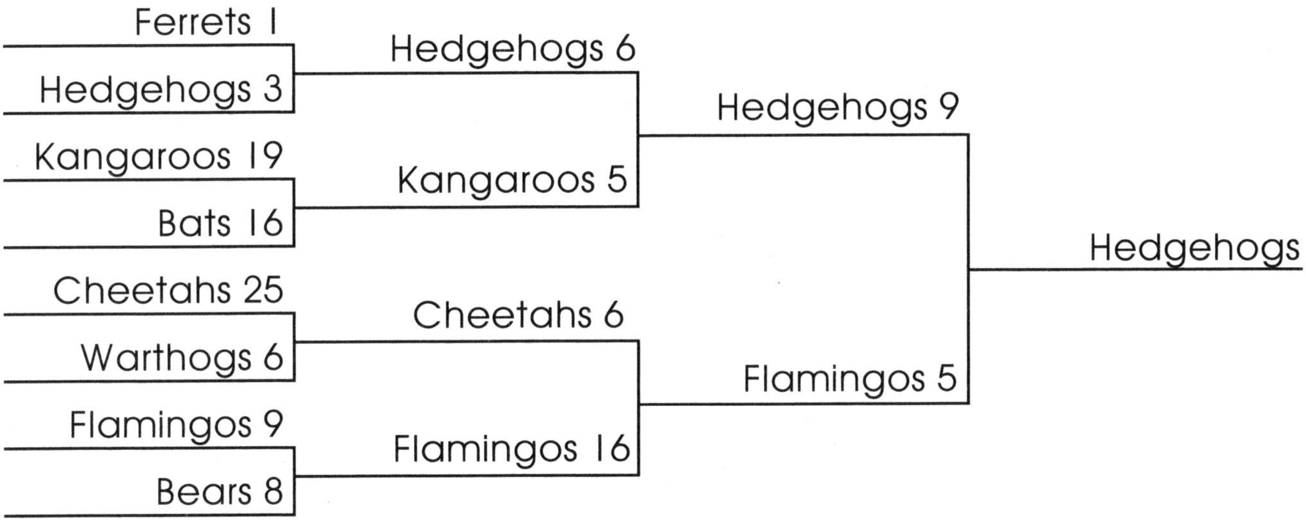

1. Which four teams won the first set of games? ____________________
2. Who was the overall winner? ____________________
3. Who had the most points for an individual game?
   team ____________________ points ____________
4. Who had the least points for an individual game?
   team ____________________ points ____________
5. What is the point difference between these two teams? ________
6. What are the median and mean scores of the following two teams?
   Hedgehogs median ____________ mean ____________
   Flamingos median ____________ mean ____________
7. Add to determine the total score for each team.
   a. Bats ______ b. Bears ______
   c. Cheetahs ______ d. Ferrets ______
   e. Flamingos ______ f. Hedgehogs ______
   g. Kangaroos ______ h. Warthogs ______
8. Did the team that had the most overall points win the tournament? Explain your answer. ____________________

____

____

 Name ____________

Game Marathon

Soa's grade level has a Friday game marathon to practice math skills. They tallied each complete game and made this graph.

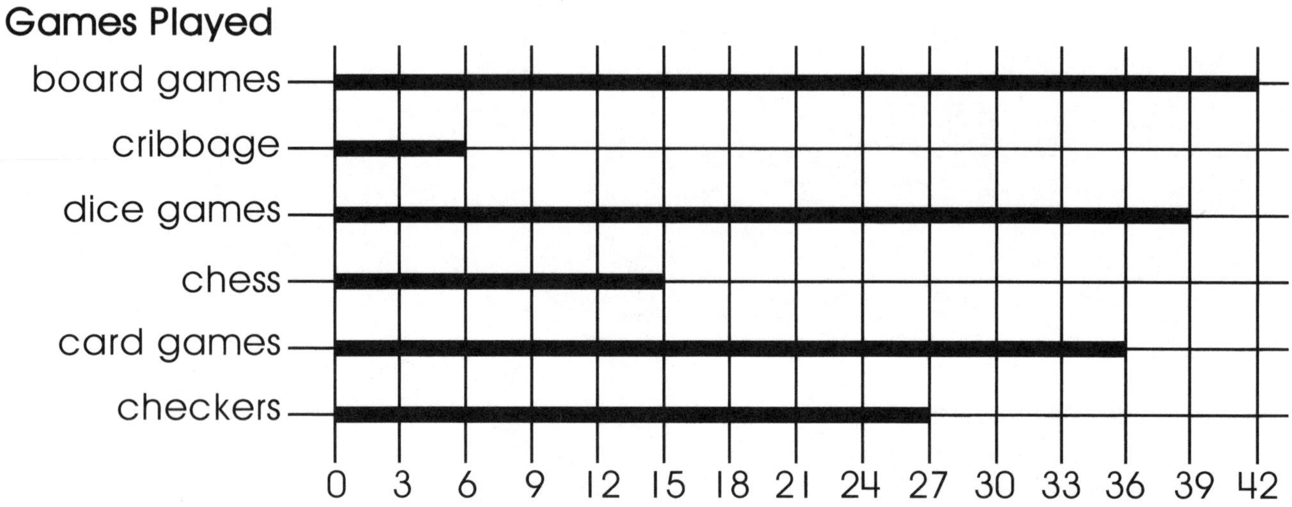

Use the graph to make a tally chart.

checkers	
card games	
chess	
dice games	
cribbage	
board games	

Use the graph to make a frequency table.

checkers	______
card games	______
chess	______
dice games	______
cribbage	______
board games	______

1. What type of games did the students play the most?

   ____

2. How many fewer games of cribbage were played than card games? ____________

3. Soa played a game of Go Fish. Where did she put her tally?

   ____

4. List the games that were played fewer than 20 times. ____________

   ____

5. How many more games of checkers were played than games of chess? ____________

Data Analysis and Probability—interpreting a graph

Name ____________________

Ladybugs

Madalen read that two-spotted ladybugs are the most common type. She wanted to know if this was true of Michigan ladybugs. Madalen collected ladybugs, counted their spots, and graphed the results. Use the graph to answer the questions.

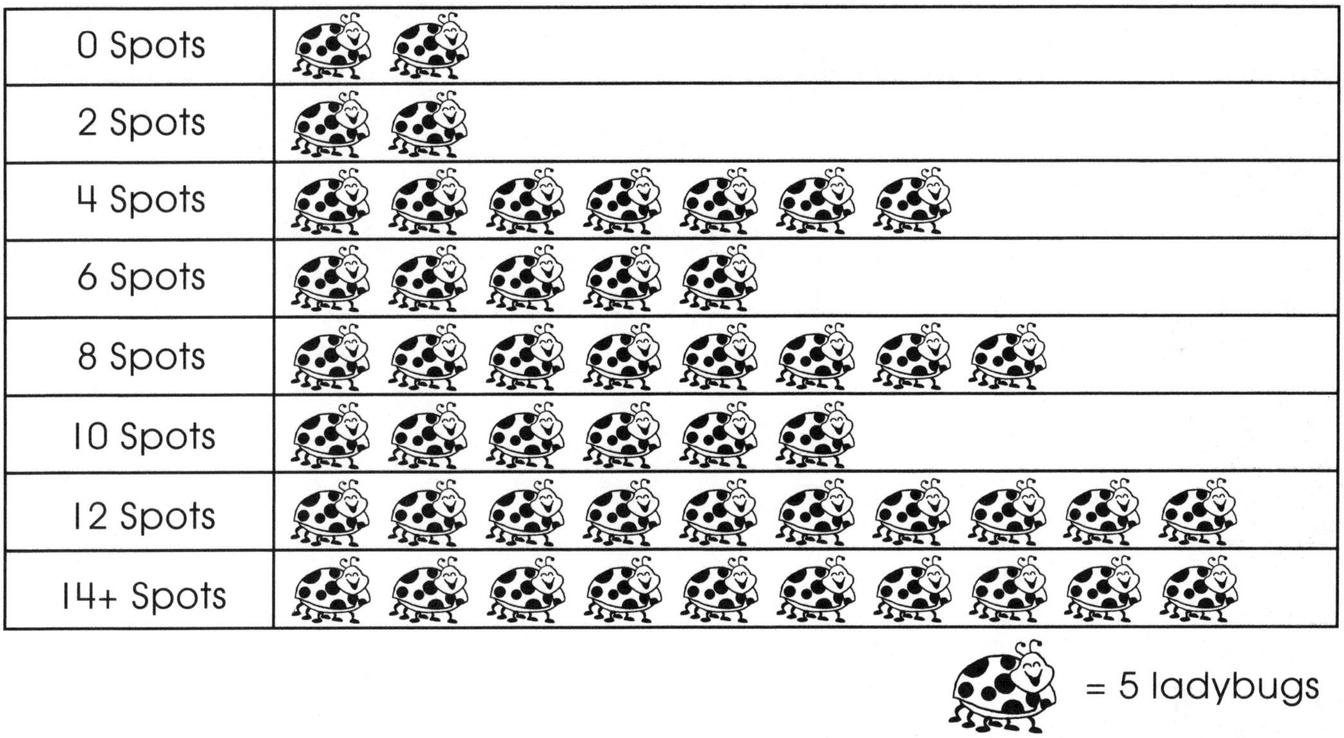

1. Madalen predicted that she would find more 2-spotted ladybugs. Was her prediction correct? Explain your answer.

   ____

2. Of which category of ladybugs did Madalen find 25? ____________

3. Of which ladybugs did Madalen find less than 20? ____________

4. Did she find more than 30 of any type of ladybug? If so, which kind? ____________________________

5. How many more 10-spotted ladybugs did she find than ones with 0 spots? ____________________________

 Name ____________________

Madalen read that 2-spotted ladybugs were the most common type. Most of the Michigan ladybugs Madalen found had more than 2 spots. Why? Solve the problems below. Write the corresponding letter with the matching answer below to find out.

1. A 3 x 9 = ______
2. P 21 ÷ 3 = ______
3. E 2 x 7 = ______
4. S 4 x 4 = ______
5. O 7 x 8 = ______
6. H 45 ÷ 9 = ______
7. H 8 x 8 = ______
8. N 8 x 4 = ______
9. L 3 x 4 = ______
10. F 6 x 4 = ______
11. W 63 ÷ 7 = ______
12. N 5 x 5 = ______
13. T 16 ÷ 8 = ______
14. B 6 x 3 = ______
15. O 48 ÷ 6 = ______
16. E 5 x 6 = ______
17. S 4 x 5 = ______
18. M 9 x 9 = ______
19. R 6 x 7 = ______
20. M 24 ÷ 8 = ______
21. A 36 ÷ 6 = ______
22. Y 28 ÷ 7 = ______
23. U 40 ÷ 4 = ______
24. G 8 ÷ 8 = ______
25. V 3 x 5 = ______
26. I 5 x 7 = ______
27. R 4 x 9 = ______
28. D 6 x 0 = ______

12 27 0 4 18 10 1 16   35 25

2 64 30   32 56 42 2 5

64 6 15 14   3 8 36 30

20 7 56 2 16   2 8

27 18 20 56 42 18   81 8 36 14

9 6 42 3 2 5   24 36 56 81

16 10 32 20 5 35 25 30 .

Name ____________________

Probable Outcomes

Probability involves whether something is likely or unlikely to happen and to what degree. If a coin were tossed, we could say that the probability of tossing heads is 1 out of 2 because there are only 2 possibilities (heads or tails), and heads is just 1 of them.

Look at the spinner. Find each probability.

1. P(7) ____________________
2. P(1) ____________________
3. P(odd number) ____________
4. P(shaded fraction) __________
5. P(6) ____________________
6. P(even number) ___________

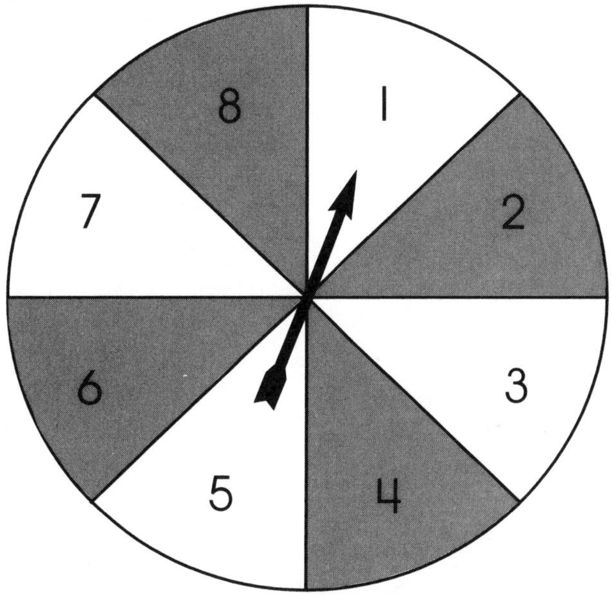

Eight scarves are placed in a hat. There is 1 blue, 2 red, 2 yellow, and 3 green. Answer the questions below.

7. What are the possible outcomes? ____________
____

8. Which 2 outcomes are equally as likely?
____

9. Which outcome is more likely than the others?
____

10. What is the probability of pulling a primary color?
____

11. What is the probability of pulling a blue or red scarf?
____

12. What is the probability of not pulling a green scarf?
____

Name ______________________

Humpback Whales

Solve the following word problems.

1. One humpback whale can eat 4,400 popcorn-size krill in 1 meal. How many krill are eaten if 8 whales have a meal? ______________________

2. A humpback's flippers can be 17 feet long. What is the total length of the 2 flippers?

   ____

3. A humpback whale will migrate 5,000 miles one way going from the summer feeding grounds to the winter breeding grounds. How many miles will it make round trip?

   ____

4. There are about 30,000 humpback whales alive today. This is 30 times more than in 1965. How many whales were alive in 1965?

   ____

5. A humpback whale can measure 40 feet long. How long would a group of 9 whales be end to end?

   ____

6. Maddie likes the monthly magazine *Whales*. A 1-year subscription is $16.32. One copy at the store is $1.76. How much would she save each month if she purchased the subscription? ______________
   How much would she save in 1 year?

   ____

7. A humpback can weigh 40 tons. This is the same as 8 full-grown male African elephants. How much does 1 elephant weigh?

   ____

Name ____________________

Movie Night

Willie wants to go to the movies. Help him solve the following problems.

1. It takes 17 minutes to get to the theater from Willie's house. The movie starts at 5:45. If Willie gives himself 5 minutes for the ticket line and 8 minutes to stand in the popcorn line, what is the latest he can leave the house and not be late for the movie?

    ____

2. Each movie that starts after 6:00 p.m. costs $7.00. All movies that start before 6:00 p.m. are $4.75. How much will Willie save if he goes to the 5:45 p.m. show?

    ____

3. Willie's movie starts at 5:45 p.m. It lasts 2 hours and 20 minutes. What time will the movie be done?

    ____

4. A candy bar costs $2.00 at the theater. The same size bar costs 69¢ at the supermarket. How much would Willie save if he bought the candy bar at the supermarket instead of the theater?

    ____

5. The theater is running 2 popcorn/soda deals. A large popcorn and soda is $6.99, a medium popcorn and soda is $4.99. Willie has $10.00. Does he have enough for a movie and a popcorn/soda deal?

    ____

Name ____________________

Theme Park

Solve the word problems below.

1. Parking lot A holds 700 automobiles. It has 528 in it. Parking lot B holds 900 automobiles. It has 417 in it. How many more cars can the 2 lots hold? ____________________

2. Use the price list to decide if the following families should buy the family pass or purchase individual tickets.

Family Pass:	$90
Adult:	$22
Child 10–16:	$18
Child 4–9:	$15
Child 0–3:	$ 9

   a. Both parents and 3 children are along in Meg's family. The children's ages are 3, 7, and 9 years old.

   ____

   b. Evita's dad is taking five 8-year-old boys.

   ____

   c. Diego's parents are taking 3 children, ages 10, 17, and 8.

   ____

3. The park opens at 10:00 a.m. and closes at 8:00 p.m. How many hours is the park open? ____________________

4. Evita's brother wore a pedometer. It says he walked 12,683 meters. How many kilometers is that? ____________________

5. One of the roller coasters is half a mile long. A mile is 5,280 feet. How many feet long is the ride? ______________
   How many yards? ______________

6. Meg bought a theme park button for $2.58 and a T-shirt for $14.73. She paid with a 20-dollar bill. What is her change? ______________
   How many park pencils can she buy with what's left over if they cost $1.33 each?

   ____

Answer Key

You've Got Value ....page 6

1. seven tens
2. seven thousands
3. nine ten thousands
4. eight hundreds
5. five ones
6. two hundreds
7. four hundreds
8. one hundred thousand
9. four hundreds
10. six hundreds
11. three tens
12. two tens
13. seven ten thousands
14. two thousands
15. six hundreds
16. seven ten thousands
17. five tens
18. seven hundreds

Ancient Creature ....page 7

1. 3,615	2. 753	3. 1,191
4. 888	5. 6,823	6. 9,642
7. 2,578	8. 1,033	9. 642
10. 1,100		

Coelacanth

That's a Lot ....page 8

1. word: two thousand one hundred fifty-six
   standard: 2,156
   expanded: 2,000 + 100 + 50 + 6
   even
2. word: three hundred nine
   standard: 309
   expanded: 300 + 9
   odd
3. word: three thousand five hundred eighty-two
   standard: 3,582
   expanded: 3,000 + 500 + 80 + 2
   even
4. word: five thousand one hundred thirty-seven
   standard: 5,137
   expanded: 5,000 + 100 + 30 + 7
   odd

Standard Form ....page 9

1. standard: 8,000,000,407
   expanded: 8,000,000,000 + 400 + 7
2. standard: 25,000
   expanded: 20,000 + 5,000
3. standard: 186,000
   expanded: 100,000 + 80,000 + 6,000
4. standard: 147,000,000
   expanded: 1,000,000,000 + 40,000,000 + 7,000,000
5. standard: 76
   expanded: 70 + 6
6. standard: 1,116
   expanded: 1,000 + 100 + 10 + 6

Comparing ....page 10

1. <	2. >	3. =	4. <
5. >	6. >	7. <	8. <
9. <	10. >	11. >	12. >
13. =	14. <	15. >	16. >
17. <	18. <	19. >	20. <

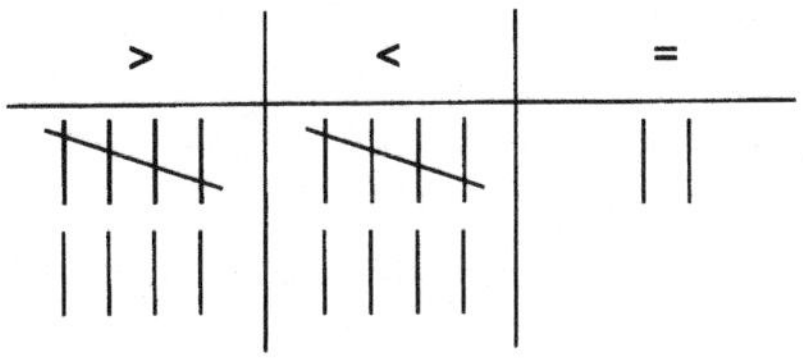

frequency table b.

Rounding...page 11

1. 30	2. 100	3. 80	4. 50
5. 80	6. 60	7. 70	8. 30
9. 20	10. 20	11. 60	12. 100
13. 40	14. 70	15. 40	16. 40
17. 50	18. 0		

Round Down:
84, 61, 33, 23, 21, 43, 74, 42, 2

Round Up:
26, 99, 75, 45, 67, 58, 98, 36, 49

If the ones digit is below 5, round down. If it is above 5, round up.

When in Rome ....page 12

1. 29	2. 70
3. 1902	4. 1400
5. 1792	6. 1752
7. 16; 6	8. 1814; 1931
9. 1998; 66; 70	

Round It ....page 13

1. 662; r10–660; r100–700
2. 765; r10–770; r100–800
3. 731; r10–730; r100–700
4. 640; r10–640; r100–600
5. 386; r10–390; r100–400
6. 583; r10–580; r100–600
7. 427; r10–430; r100–400
8. 505; r10–510; r100–500
9. 633; r10–630; r100–600
10. 358; r10–360; r100–400
11. 542; r10–540; r100–500
12. 371; r10–370; r100–400
13. 228; r10–230; r100–200
14. 903; r10–900; r100–900
15. 771; r10–770; r100–800

Estimate v. Actual ....page 14

1.	50	2.	90
3.	260	4.	570
5.	10	6.	40
7.	630	8.	160
9.	300	10.	800
11.	400	12.	400
13.	600	14.	500
15.	800	16.	100
17.	488; 400	18.	861; 900
19.	388; 400	20.	679; 700
21.	288; 300	22.	1,113; 1,000
23.	823, 800	24.	511; 500

Back and Forth.....................................page 15

1. largest = 763; smallest = 367
   763 – 367 = 396; 763 + 367 = 1,130;
   367 + 763 = 1,130
2. largest = 853; smallest = 358
   853 – 358 = 495; 853 + 358 = 1,211;
   358 + 853 = 1,211
3. largest = 731; smallest = 137
   731 – 137 = 594; 731 + 137 = 868;
   137 + 731 =868
4. largest = 972; smallest = 279
   972 – 279 = 693; 972 + 279 = 1,251;
   279 + 972 = 1,251
5. largest = 741; smallest = 147
   741 – 147 = 594; 741 + 147 = 888;
   147 + 741 = 888
6. largest = 995; smallest = 599
   995 – 599 = 396; 995 + 599 = 1,594;
   599 + 995 = 1,594
7. largest = 511; smallest = 115
   511 – 115 = 396, 511 + 115 = 626;
   11 5 + 551 = 626
8. largest = 852; smallest = 258
   852 – 258 = 594; 852 + 258 = 1,110;
   258 + 852 = 1,110
9. largest = 651; smallest = 156
   651 – 156 = 495; 651 + 156 = 807;
   156 + 651 = 807

Kaleidoscope ....page 16

1.
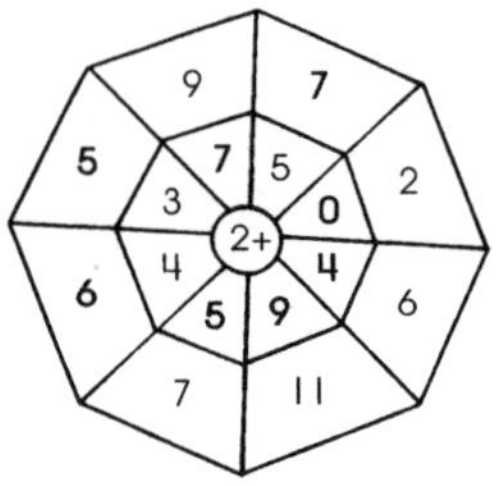

2.
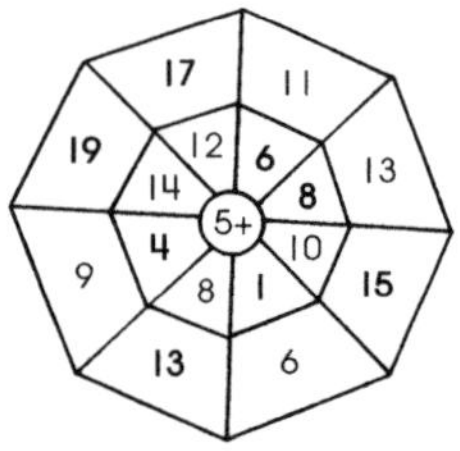

3.
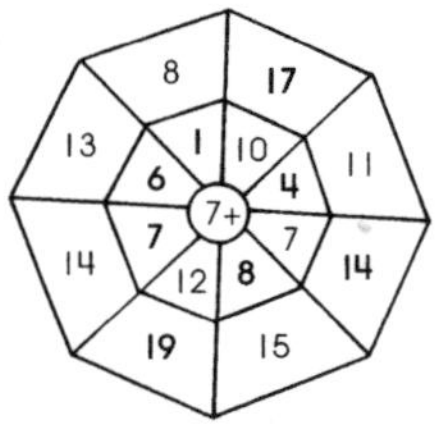

4.
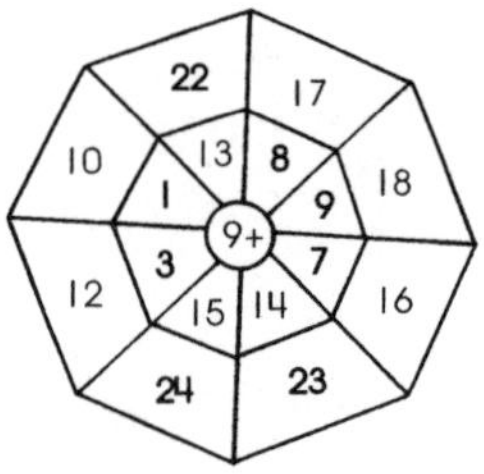

5.
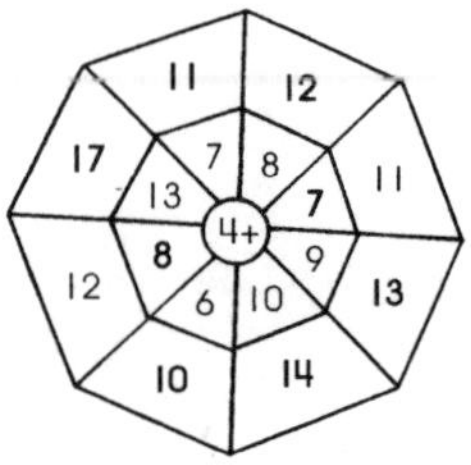

6. even
7. odd
8. odd
9. even

Addition Boxes .... page 17

1. 123 2. 121 3. 92 4. 56
5. 113 6. 80 7. 73 8. 111
9. 770 10. 825 11. 573 12. 904
13. 532 14. 824 15. 823

American States .... page 18

1. 5,009 2. 9,304
3. 2,057 4. 7,836
5. 6,450 6. 1,214
7. 8,257 8. 9,609
9. 1,819 10. 1,788
11. 1,959 12. 1,912
13. 1,850 14. 1,907
15. 1,777 16. 1,796

Picture This .... page 19

1. 35 2. 22 3. 15 4. 48
5. 43 6. 18 7. 38 8. 36
9. 33 10. 15 11. 66 12. 16
13. 7 14. 18

Mount Rushmore .... page 20

1. 5,043 2. 6,135 3. 8,117
4. 2,727 5. 4,683 6. 7,740
7. 5,298 8. 9,201 9. 8,228
10. 6,823 11. 7,803 12. 7,931
13. 4,217 14. 7,525 15. 10,125

Thomas Jefferson, Abraham Lincoln, Theodore Roosevelt, George Washington

Climbing .... page 21

Across

1. 9,417 2. 5,321
3. 8,901 4. 8,436
5. 9,262 6. 3,622
7. 5,027

Down

2. standard: 5,436
   expanded: 5,000 + 400 + 30 + 6
3. standard: 8,310
   expanded: 8,000 + 300 + 10
4. standard: 8,927
   expanded: 8,000 + 900 + 20 + 7
5. standard: 9,401
   expanded: 9,000 + 400 + 1
6. standard: 3,231
   expanded: 3,000 + 200 + 30 + 1
7. standard: 5,666
   expanded: 5,000 + 600 + 60 + 6
8. standard: 7,022
   expanded: 7,000 + 20 + 2

What's Your Score? .... page 22

1. 42 – 26 = 16 2. 72 + 58 = 130
3. 29 + 34 = 63 4. 78 – 49 = 29
5. 65 – 37 = 28 6. 14 + 37 = 51
7. 59 – 16 = 43 8. 55 + 46 = 101
9. 91 + 28 = 119 10. 43 – 27 = 16
11. 80 – 63 = 17 12. 81 – 56 = 25
13. 49 + 38 = 87 14. 69 + 45 = 114

C = 100%

Rapunzel .... page 23

1. < 2. > 3. < 4. =
5. = 6. > 7. = 8. <
9. > 10. < 11. = 12. =
13. < 14. < 15. > 16. >
17. < 18. > 19. < 20. <

Left Rapunzel

Table Time .... page 24

x	1	2	3	4	5	6	7	8	9
1	1	2	3	4	5	6	7	8	9
2	2	4	6	8	10	12	14	16	18
3	3	6	9	12	15	18	21	24	27
4	4	8	12	16	20	24	28	32	36
5	5	10	15	20	25	30	35	40	45
6	6	12	18	24	30	36	42	48	54
7	7	14	21	28	35	42	49	56	63
8	8	16	24	32	40	48	56	64	72
9	9	18	27	36	45	63	54	72	81

1. 9 2. 3 3. 5 4. 2
5. 9 6. 6 7. 4 8. 5
9. 7 10. 5 11. 9 12. 8
13. 3 14. 6 15. 5 16. 8
17. 7 18. 6 19. 7 20. 5
21. 4 22. 8 23. 1 24. 9
25. 5

Order Up! .... page 25

1. 38
2. 19.65
3. 27.94
4. 38, 27.94, 10.65
5. 11.75
6. 4.6
7. 6.38
8. 11.75, 6.38, 4.6
9. 6.36
10. 7.55
11. 4.61
12. 7.55, 6.36, 4.61

Ssssnakes .... page 26

1. 286.75	2. 82.94	3. 582.52
4. 164.76	5. 40.31	6. 44.66
7. 1.87	8. 87.93	9. 179.48
10. 180.51	11. 142.68	12. 47.77
13. 257.11	14. 65.84	15. 3.64

Snake #2

Multiplication .... page 27

1. three rockets with five windows on each
   3 x 5 = 15
2. five bears with two eyes on each
   5 x 2 = 10
3. four pockets with eight marbles in each
   4 x 8 = 32
4. two fish bowls with six fish in each
   2 x 6 = 12
5. five octopuses with eight arms on each
   5 x 8 = 40
6. one robot with nine circular parts
   1 x 9 = 9

Traveling .... page 28

1. 469—**c.** 459	2. 285—**h.** 225
3. 873—**e.** 243	4. 351—**f.** 301
5. 448—**a.** 738	6. 196—**b.** 126
7. 180—**i.** 150	8. 294—**j.** 224
9. 292—**d.** 252	10. 57—**g.** 77

Cobwebs .... page 29

1.

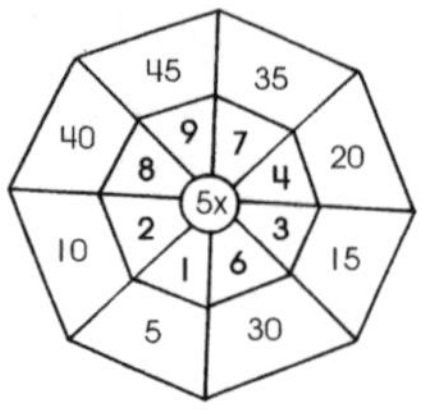

2.

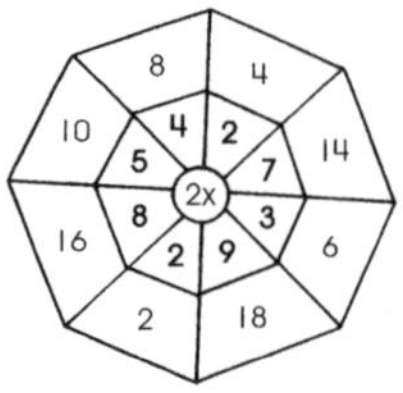

3.

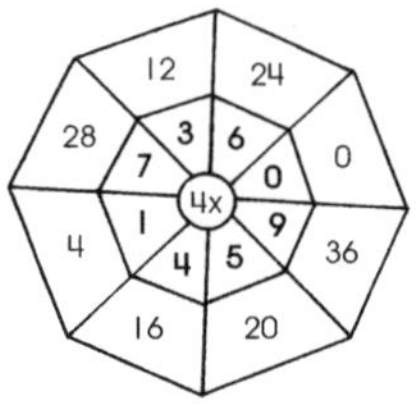

4.

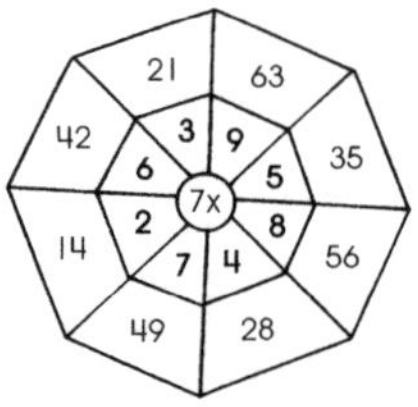

5.

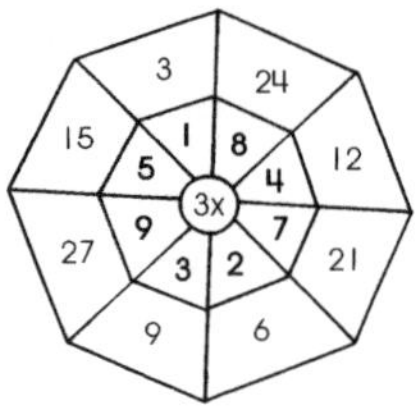

All Boxed Up .... page 30

1. 576	2. 245	3. 1,054
4. 364	5. 192	6. 1,752
7. 432	8. 444	9. 4,332
10. 51	11. 1,370	12. 2,674

For the Romans .... page 31

1. 318; d	2. 423; a	3. 256; l	4. 105; i
5. 272; f	6. 518; c	7. 117; e	8. 162; g
9. 360; h	10. 588; b		

Two-Digit Place Value Boxes .... page 32

1. 3, 276	2. 2,117	3. 1,404
4. 736	5. 714	6. 5,740
7. 2,015	8. 3,648	

Multiplying Up ....page 33

1.

120

12	10

6	2	5

2.

384

16	24

2	8	3

3.

567

21	27

7	3	9

4.

240

20	12

5	4	3

5.

216

18	12

9	2	6

6.

1,536

48	32

6	8	4

7.

288

12	24

2	6	4

8.

675

15	45

3	5	9

9.

270

18	15

6	3	5

2	3	1	5

10.

864

72	12

12	6	2

4	3	2	1

11.

640

20	32

5	4	8

5	1	4	2

12.

3,072

48	64

6	8	8

3	2	4	2

3	1	2	2	1

13.

1,920

40	48

10	4	12

5	2	2	6

5	1	2	1	6

Division ....page 34

1. three rockets with five windows on each
   15 ÷ 3 = 5
2. five bears with two eyes on each
   10 ÷ 5 = 2
3. four pockets with eight marbles in each
   32 ÷ 4 = 8
4. two fish bowls with six fish in each
   12 ÷ 2 = 6
5. five octopuses with eight arms on each
   40 ÷ 5 = 8
6. one robot with nine circular parts
   9 ÷ 1 = 9

Table Tool..page 35

+	1	2	3	4	5	6	7	8	9
1	2	3	4	5	6	7	8	9	10
2	3	4	5	6	7	8	9	10	11
3	4	5	6	7	8	9	10	11	12
4	5	6	7	8	9	10	11	12	13
5	6	7	8	9	10	11	12	13	14
6	7	8	9	10	11	12	13	14	15
7	8	9	10	11	12	13	14	15	16
8	9	10	11	12	13	14	15	16	17
9	10	11	12	13	14	15	16	17	18

1. 4 2. 6 3. 9 4. 4
5. 6 6. 4 7. 9 8. 2
9. 6 10. 8 11. 7 12. 9
13. 7 14. 1 15. 8 16. 4

17.	5	**18.**	8	**19.**	9	**20.**	2
21.	7	**22.**	1	**23.**	7	**24.**	6

Schooling ....page 36

1.	27	**2.**	46	**3.**	55	**4.**	128
5.	38	**6.**	69	**7.**	71	**8.**	178
9.	18	**10.**	99	**11.**	24	**12.**	141

Practice, Practice ....page 37

1.	13	**2.**	1	**3.**	7	**4.**	13
5.	1	**6.**	9	**7.**	19	**8.**	13
9.	5	**10.**	12	**11.**	20	**12.**	1
13.	4	**14.**	21	**15.**	14		

Mary Anning ....page 38

1.	3	**2.**	24	**3.**	12	**4.**	9
5.	2	**6.**	6	**7.**	49	**8.**	7
9.	15	**10.**	1	**11.**	9	**12.**	5
13.	8	**14.**	4	**15.**	20	**16.**	28
17.	36	**18.**	32				

19.	756	She
20.	148	sells
21.	389	seashells
22.	561	by
23.	278	the
24.	610	seashore.

Missing Numbers ....page 39

Puzzle One: 48, 8, 4, 36, 6, 18, 2, 7, 28, 14
Puzzle Two: 9, 9, 8, 4, 24, 3, 5, 7, 8, 8
Puzzle Three: 5, 9, 6, 12, 3, 7, 4, 20, 10, 30

Serving Size ....page 40-41

1. 3; 210 calories
2. 2; 180 calories
3. 2; 210 calories
4. 3; 510 calories
5. 7; 742 calories
6. 11; 448 calories
7. 924 calories
8. 932 calories
9. 1,305 calories; milk—6; popcorn—4 1/2
10. 427 calories
11. 690 calories
12. 478 calories
13. 658 1/2 calories; 6 1/4 servings
14. 594 calories

Fractions...page 42

1. **a.** 1/5; **b.** 1/3; **c.** 2/7; **d.** 3/9 = 1/3; **e.** 1/8; **f.** 2/4= 1/2; **g.** 1/6; **h.** 3/10
2. **a.** 2/5; **b.** 1/3; **c.** 3/7; **d.** 1/9; **e.** 4/8 = 1/2; **f.** 1/4; **g.** 2/6= 1/3; **h.** 6/10= 4/5
3. **a.** 2/5; **b.** 1/3; **c.** 2/7; **d.** 5/9; **e.** 3/8; **f.** 1/4; **g.** 3/6= 1/2; **h.** 1/10

Pull Apart ....page 43

1. 4; **a.** 1/4; **b.** 1/4 **c.** 2/4 = 1/2; **d.** 3/4; **e.** 4/4 = 1
2. 3; **a.** 1/3; **b.** 1/3; **c.** 1/3; **d.** 2/3; **e.** 2/3
3. 7; **a.** 2/7; **b.** 1/7; **c.** 3/7; **d.** 1/7; **e.** 4/7; **f.** 3/7
4. 5; **a.** 1/5; **b.** 2/5; **c.** 2/5; **d.** 3/5; **e.** 4/5; **f.** 5/5 = 1

Drawing Fractions ....page 44

Answers will vary but should reflect given fraction.

Going Fishing ....page 45

1.	4/14	**2.**	1/14	**3.**	3/14
4.	3/14	**5.**	1/14	**6.**	2/14
7.	2/14	**8.**	7/14	**9.**	5/14

10. 9/14
11. a and c
12. Answers will vary but should equal 7.

Circles, Squares, and ... .page 46

1. 3 circles; 2 are blue.
2. 2 triangles; 1 is purple.
3. 8 rectangles; 5 are yellow.
4. 5 ovals; 2 are green and 3 are yellow.
5. 4 circles; 1 is green, 2 are blue, and 1 is purple.
6. 7 squares; 2 are brown, 4 are red, and 1 is white.
7. 4 rectangles; 2 are red.
8. 8 squares; 6 are blue.

Pizza Pie ....page 47

1. 2/4, 3/6, 8/16; 1/2 is colored.
2. 4/6, 10/15, 6/9, 14/21; 2/3 is colored.
3. 9/12, 18/24, 15/20, 27/36; 2/3 is colored.
4. 9/15, 12/20, 6/10; 1/2 is colored.

Answers will vary.

Fractionality ....page 48

1. **blue:** 1/3, 4/9, 2/10, 7/20, 1/5, 1/8, 3/7, 4/12
red: 2/4, 3/6, 6/12, 1/2, 4/8, 5/10, 8/16, 7/14, 12/24, 6/12, 3/6, 4/8, 2/4, 10/20, 8/16, 50/100
yellow: 5/7, 8/10, 11/12, 8/9, 99/100, 4/5, 3/4, 17/19, 40/50, 3/4, 6/9, 16/20, 7/9, 4/7, 4/6, 8/10, 7/9, 14/15, 70/100, 17/18, 5/8, 12/13, 2/3,

6/7, 4/5, 2/3, 6/10, 9/10, 7/10, 5/7, 6/8, 4/6, 5/9, 7/8, 5/6, 6/7

2. **blue:** 3/3, 2/2, 10/10, 4/4, 7/7, 2/2, 5/5, 6/6, 3/3, 11/11, 6/6, 8/8, 7/7, 35/35, 1/1, 6/6, 4/4, 9/9, 16/16, 5/5

   red: 5/4, 16/10, 8/3, 16/7

   yellow: 4/5, 3/6, 6/10, 2/7, 3/6, 4/9, 2/5, 3/7, 1/13, 1/9, 1/3, 7/11, 3/5, 3/10, 1/6, 1/2, 2/6, 3/7, 11/2, 1/9, 1/7, 1/5, 4/7, 7/9, 3/4, 4/6

You did it!

Capture the Flag ....page 49

1. 1/3 **2.** 1/10 **3.** 2/4 **4.** 1/6
5. 1/9 **6.** 6/18 **7.** 1/9 **8.** 1/3
9. 2/6 **10.** 1/7 **11.** 2/5 **12.** 1/9
13. 1/3 **14.** 1/2 **15.** 3/4 **16.** 1/2
17. 1/4 **18.** 6/8 **19.** 1/3 **20.** 3/5
21. 2/3 **22.** 1/7 **23.** 7/9 **24.** 9/10

Jade captures the flag.

Fractional Order ....page 50

1. 0/4, 1/4, 2/4, 3/4, 4/4; This is the correct order.
2. 1/8, 1/6, 1/5, 1/3; You got this one too.
3. 1/8, 3/8, 4/8, 6/8, 7/8; You bet. Way to go.
4. 1/16, 1/12, 1/4, 1/2, 1/1; Not this one. Just kidding.
5. 2/6, 3/6, 4/6, 5/6, 6/6; You are good at fractions.

Part Of ....page 51

1. **a.** 4; **b.** 3; **c.** 2; **d.** 6; **e.** 15
2. **a.** 6; **b.** 11; **c.** 9; **d.** 2; **e.** 28
3. **a.** 3; **b.** 9; **c.** 2; **d.** 6; **e.** 20
4. **a.** 6; **b.** 16; **c.** 10; **d.** 4; **e.** 36
5. **a.** 8; **b.** 20; **c.** 16; **d.** 14; **e.** 58
6. **a.** 8; **b.** 32; **c.** 20; **d.** 16; **e.** 66

7. 16 **8.** 21 **9.** 16 **10.** 12
11. 24 **12.** 10

Around the World ....page 52

1. **a.** 12; **b.** 30; **c.** 27; **d.** 9; Asia
2. **a.** 4; **b.** 11; **c.** 15; **d.** 8; South America
3. **a.** 6; **b.** 24; **c.** 4; **d.** 33; Australia
4. **a.** 6; **b.** 20; **c.** 8; **d.** 9; Antarctica

Adding Fractions..................................page 53

1. 3/5 **2.** 2/3 **3.** 8/10 **4.** 7/8
5. 5/12 **6.** 7/7 **7.** 5/7 **8.** 9/11
9. 4/5 **10.** 9/12 **11.** 4/6 **12.** 7/10
13. 3/8 **14.** 11/12

Subtracting Fractions ....page 54

1. 1/9 **2.** 2/3 **3.** 7/10 **4.** 1/12
5. 2/7 **6.** 1/4 **7.** 4/8 **8.** 3/5
9. 1/12 **10.** 3/7 **11.** 4/7 **12.** 2/11
13. 2/8 **14.** 1/6 **15.** 5/10 **16.** 4/8

How Ya Doing? ....page 55

1. 1/4 (4, 1) **2.** 3/5 (5, 3)
3. 4/6 (6, 4) **4.** 2/9 (9, 2)
5. 5/10 (10, 5) **6.** 2/4 (4, 2)
7. 6/7 (7, 6) **8.** 4/5 (5, 4)
9. 4/8 (8, 4) **10.** 5/11 (11, 5)
11. 2/7 (7, 2) **12.** 6/12 (12, 6)
13. 5/6 (6, 5) **14.** 2/11 (11, 2)
15. 1/6 (6, 1) **16.** 1/3 (3, 1)
17. 5/9 (9, 5) **18.** 5/12 (12, 5)
19. 3/7 (7, 3) **20.** 6/10 (10, 6)

You are doing a great job.

Decimals ....page 56

1. 0.08 **2.** 6.12 **3.** 31.7 **4.** 11.71
5. 9.09 **6.** 52.7 **7.** 0.48 **8.** 3.81
9. 0.45 **10.** 21.3 **11.** 0.56 **12.** 45.02
13. 98.11 **14.** 0.1 **15.** 0.63 **16.** 0.07
17. 0.83 **18.** 16.5

Slaying Dragons ....page 57

1. 1/10; 10% **2.** 23%; 0.23
3. 67/100; 0.67 **4.** 5/100; 0.05
5. 40%; 0.4 **6.** 9/100; 9%
7. 31/100; 31% **8.** 2%; 0.02

Pocket Change ....page 58

1. $1.40
2. $1.24
3. $1.05
4. #2 and #3
5. 12 dimes; 2 pennies
6. Answers will vary.
7. Answers will vary.

Money Riddles..................................page 59

1. I quarter, I dime, three nickels
2. 10 nickels, 2 pennies
3. I quarter, I dime, I penny
4. 2 half-dollars, I quarter, I dime, 3 nickels
5. 2 quarters, 2 dimes, 4 pennies
6. 2 dimes, 6 nickels
7. I half-dollar, I quarter, I dime, 2 pennies
8. Answers will vary.

Spending Money ....page 60

1. 37¢; 3 dimes, I nickel, 2 pennies
2. 37¢; I quarter, I dime, 2 pennies
3. $1.92; 6 quarters; 4 dimes; 2 pennies
4. $1.14; 2 half-dollars, 2 nickels, 4 pennies
5. $3.30; 5 half-dollars, 2 quarters, 3 dimes
6. $1.90; 3 half-dollars, 2 dimes, 4 nickels

Acorns ....page 61

1. $7.65
2. $7.65
3. $7.00
4. $9.05
5. $6.25

Book Order..page 62

1. $6.11
2. yes; 19¢ left over
3. 11 quarters
4. 36¢ less than estimated
5. $10.00—estimated; $9.72—actual; 28¢ difference
6. $1.59

Post Office ....page 63

1. 24 **2.** 7 **3.** 14 **4.** 6

Pia and her family of 6 are going on a 14-day vacation. They are having the post office hold their mail from June 24 to July 7.

5. 6 **6.** 5 **7.** 7 **8.** 1

Jade needs 6 stamps. Her sister Mikaela needs 5, which is one less than Jade needs. How many stamps do they need in all? 11 Each stamp costs 7¢. What is the total cost of the stamps? 77¢ Jade has a 1 dollar bill. How much change will she get back? 23¢

9. 30 **10.** 28 **11.** 16 **12.** 2

13. 10

Aaron has to mail some packages. The largest weighs 30 ounces, the middle-size package weighs 28 ounces, and the smallest weighs 16 ounces. Each weight is a multiple of 2. It costs 10¢ per ounce to send the packages. What is the total weight of the packages? 74 ounces How much will it cost? $7.40 Aaron has a 10 dollar bill. What is his change back? $2.60

Money, Money 64–65

1. $6.37 **2.** $4.96 **3.** $5.35 **4.** $7.51

5. $6.89

Ana—$4.96
Eva—$6.89
Ruy—$7.51
Sep—$5.35
Ziv—$6.37

Farmer's Market ....page 66

apples—$41.36; 64¢
avocados—$8.69; 31¢
blueberries—$24.79; 21¢
broccoli—$7.28; 72¢
cantaloupe—$12.45; 55¢
carrots—$9.18; 82¢
cucumbers—$2.88; $2.12
green beans—$13.92; 8¢
peaches—$20.58; 42¢
tomatoes—$10.32; 18¢
strawberries—$39.99; 1¢

1.–2. Answers will vary.

Envisioning ....page 67

0, 1, 2, 3, 4, 5, 6, 7, 8, 9
Answers will vary.

Left to Right ....page 68

Answers will vary. Sample answers are:

1. 16 = 8 x 2; 2 x 2 = 4 x 1; 4 + 1 = 5; 4 + 1 + 5 = 10
2. 10 + 7 + 3 < 21; 5 + 2 + 9 = 3 x 3 + 7; 9 = 3 x 3; 7 x 1 = 7
3. 17 = 1 + 2 x 8; 2 + 8 = 10; 3 + 6 – 1 = 8; 8 + 7 > 6 + 5
4. 0 x 15 = 0 x 5; 5 + 3 = 7 – 1; 3 + 7 – 1 = 4 + 5; 4 + 5 = 9 + 0 x 6
5. 6 = 1 x 6; 6 x 3 = 18; 18 = 9 + 9; 0 + 7 x 1 < 8

Going to the Top ....page 69

1.

39
19 20
10 9 11
4 6 3 8

2.

41
21 20
12 9 11
5 7 2 9

3.

50
20 30
7 13 17
3 4 9 8

4.

119
56 63
22 34 29
7 15 19 10
2 5 10 9 1

5.

96
48 48
24 24 24
11 13 11 13
4 7 6 5 8

6.

335
155 180
74 81 99
38 36 45 54
20 18 18 27 27
9 11 7 11 16 11
3 6 5 2 9 7 4

Brick Work..page 70

1. 3	**2.** 6	**3.** 7, 8	**4.** 2, 8
5. 4, 0	**6.** 0, 9	**7.** 4, 9	**8.** 7, 5
9. 4	**10.** 5, 3	**11.** 1, 0	**12.** 5, 4
13. 1, 9	**14.** 2, 8	**15.** 4, 7	**16.** 2, 7
17. 4, 5	**18.** 1, 2	**19.** 2, 3	**20.** 7, 6

Mason

Give Me an N ....page 71

1. f = 20
2. m = 8
3. y = 9
4. h = 36
5. c = 7
6. j = 13; n = 5
7. i = 3; v = 2
8. g = 8; w = 10; y = 2
9. f = 3, h = 6, m = 10, p = 5

Number Letters ....page 72

1. 4	**2.** 40	**3.** 6	**4.** 8
5. 9	**6.** 3	**7.** 10	**8.** 21
9. 12	**10.** 7	**11.** 5	**12.** 2
13. 1	**14.** 18	**15.** 27	

Number letters don't scare me!

Order Up ....page 73

1. pencil, marker, chalk, crayon, pen
2. bird, mouse, cat, dog, ant
3. milk, orange juice, water, cola, root beer
4. apple, grapes, banana, orange, pear
5. hexagon, triangle, circle, quadrilateral, octagon

Scavenger Hunt ....page 74

1. (2, 3)	**2.** (3, 3)	**3.** (5, 4)	**4.** (3, 5)
5. (6, 2)	**6.** (4, 2)	**7.** (5, 7)	**8.** (0, 4)
9. (2, 6)	**10.** (1, 7)	**11.** (1, 4)	**12.** (6, 6)
13. (5, 1)	**14.** (6, 3)	**15.** (3, 2)	

Are You Included? ....page 75

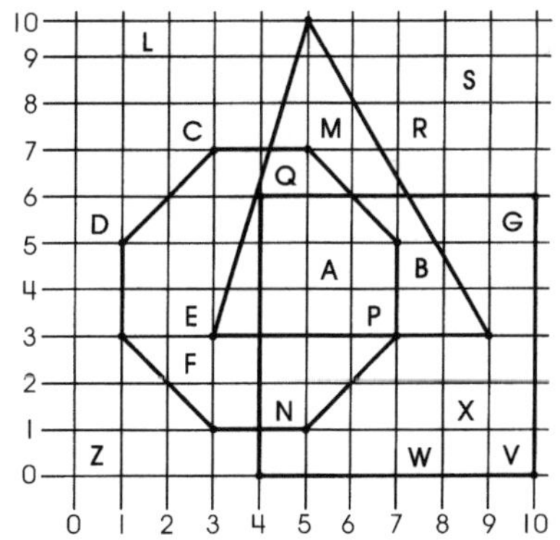

1. N, W, X, V, G
2. A, P, B
3. E, F
4. Q, A, P

5. C, Q, M, A, P, B, R, S
6. none
7. C, R, S, E, F
8. A, P
9. L, Z

Are We the Same? .... **page 76**

1. yes 2. no 3. yes 4. yes
5. no 6. no

Answers will vary.

Describe Each Figure .... **page 77**

1. sides: 3; vertices: 3
2. sides: 7; vertices: 7
3. sides: 4; vertices: 4
4. sides: 12; vertices: 12
5. sides: 9; vertices: 9
6. sides: 6; vertices: 9

Match Up .... **page 78**

1. polygon

2. rectangle

3. octagon

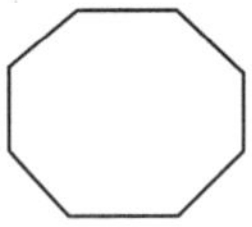

4. triangle

5. trapezoid

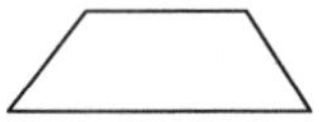

6. pentagon

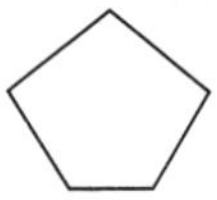

7. parallelogram

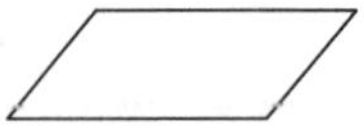

8. hexagon

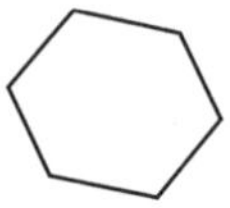

9. square

Quadrilaterals .... **page 79**

X on all 4 triangles, 2 pentagons, 3 octagons, and 2 hexagons
3 parallelograms circled
5 rectangles blue
4 trapezoids green
4 squares red

What's Your Angle .... **page 80**

1. right 2. acute
3. acute 4. acute
5. acute 6. acute
7. acute 8. obtuse
9. obtuse 10. obtuse
11. obtuse 12. obtuse
13. obtuse 14. obtuse

Solids .... **page 81**

Cone	1	0	0
Rectangular prism	6	12	8
Cylinder	2	0	0
Octagonal prism	10	24	16
Cube	6	12	8
Triangular prism	5	9	6
Hemisphere	1	0	0
Sphere	0	0	0
Square pyramid	5	8	5

1. sphere
2. triangular prism
3. cone
4. cube
5. cylinder

Dimensions .... **page 82**

1. triangle, circle
2. square, triangle
3. rectangle
4. circle, rectangle
5. 6 squares
6. 5 triangles
7. 2 octagons, 8 rectangles
8. circle

Around the Rim .... **page 83**

1. 70 in.
2. 36 yd.
3. 52 m
4. 128 cm

5. 84 ft.
6. 32 cm
7. 66 m
8. 120 in.

Tiling ....page 84

1. perimeter: 8 units; area: 3 sq. units
2. perimeter: 12 units; area: 8 sq. units
3. perimeter: 14 units; area: 10 sq. units
4. perimeter: 22 units; area: 13 sq. units
5. perimeter: 24 units; area: 22 sq. units
6. perimeter: 18 units; area: 12 sq. units
7. perimeter: 12 units; area: 8 sq. units

Calculating Area ....page 85

1. area: 36 sq. units; perimeter: 26 units
2. area: 50 sq. units; perimeter: 30 units
3. area: 48 sq. units; perimeter: 28 units
4. area: 36 sq. units; perimeter: 30 units
5. area: 49 sq. units; perimeter: 28 units
6. area: 80 sq. units; perimeter: 36 units
7. 32 sq. inches
8. 30 sq. inches

Cover It ....page 86

1. 693 sq. units
2. 304 sq. units
3. 2,088 sq. units
4. 5,037 sq. units
5. 4,180 sq. units
6. 4,446 sq. units

>2,000: 2,088; 4,180; 4,446
odd: 693
both: 5,037
outside set: 304

Shadows..page 87

1. 21 sq. units
2. 64 sq. units
3. 24 sq. units
4. 18 sq. units
5. 24 sq. units
6. 15 sq. units

Fill It Up ....page 88

1. 18 cubic units
2. 16 cubic units
3. 15 cubic units
4. 40 cubic units
5. 36 cubic units
6. 32 cubic units

Upping the Volume ....page 89

1. 360 cubic units
2. 1,435 cubic units
3. 384 cubic units
4. 1,456 cubic units
5. 612 cubic units
6. 348 cubic units
7. 432 cubic feet of cement
8. 9 x 9 x 2 pan

Wall of Snow ....page 90

Jake: 6 x 6 x 6
Jen: 8 x 4 x 12
Jill: 12 x 12 x 6
Josh: 12 x 12 x 12
Juri: 12 x 6 x 8

1. 96 cubic blocks
2. 99 cubic blocks
3. 32 cubic blocks
4. 15 cubic blocks

From Here to There ....page 91

36 inches
5,280 feet

1. 4 ft. 4 in.
2. 5,280 yd.; 15,840 ft.
3. 818 yd.
4. 84 in.
5. 4 yd. 1 ft.
6. no
7. 13 yd.
8. 72 in.
9. 6 boards

Measure This ....page 92

1. 7 cm
2. 5 cm
3. 16 cm
4. 2 cm
5. 13 cm
6. ¾ in.
7. 5 ¼ in.
8. 3 ½ in.
9. 2 ¼ in.

10. 4 ¾ in.
11.–15. Answers will vary.

Metric Length ....page 93

100 centimeters; 1,000 millimeters;
100 decimeters; 1,000 centimeters;
10 decameters; 10 hectometers

1. meters or decameters; too long for decimeters and too short for hectometers
2. decimeters or centimeters
3. 345 millimeters; 3.45 decimeters
4. 158 decameters; 15.8 hectometers
5. yes
6. 127 centimeters; 12.7 decimeters; 1,270 millimeters

Tonnage ....page 94

1. 80 ounces
2. 4 bags
3. 78,000 pounds
4. 3 tons
5. 532 kg
6. 1,214 g wheat; 1,352 g oats
7. 195 mg; 805 mg left over

Measuring Mass...................................page 95

1. actual mass: 6 g; actual mass is 2 less than the estimated of 8 g
2. The actual mass of 86 grams is 46 more grams than the estimated mass of 40.
3. 15 g; 5 g for each balloon
4. The actual mass is 9 more grams than the estimated.
5. 4 g per clip
6. 1 gram
7. 61 g
8. 66 g
9. 28 g

Capacity ....page 96

48 tsp., 16 fl. oz., 4 c., 32 fl. oz., 16 c., 8 pt., 128 fl. oz.

1. no
2. 60¢; $1.20
3. 80 drops
4. 48 tbsp.; 144 tsp.
5. 10 cups; no

Metric Capacity.................................page 97

1. 10 mL
2. 3 tsp.
3. 480 mL; no, too much
4. 23,600 mm^3; 236 dm^3
5. 6.4 kL
6. 20 bottles

Brrrr, It's Cold! ....page 98

1. 70° F; 22° C
2. 32° F; 0° C
3. 90° F; 32° C
4. 50° F; 10°C
5. 15°F; -10° C
6. 60° F; 15° C

What Time? ....page 99

1. 3:50; 4:50
2. 9:15; 8:15
3. 2:10; 1:50
4. 7:30; 8:25
5. 5:20; 4:35
6. 6:45; 7:50
7. 8:05; 6:05
8. 12:10; 11:20

On Time ....page 100

1. She will be late by 5 minutes.
2. He must leave by 7:50.
3. She can play for 20 more minutes.
4. She will get home at 7:50.
5. No, he will miss 15 minutes of the show.
6. He must start at 12:25.

Have the Time?..................................page 101

1. 55 minutes; he must get up by 7:25.
2. 240 seconds are in 4 minutes; she has 205 seconds; she can make about 8 trades.
3. 1 hour and 45 minutes, 70 minutes left; they take him 64 minutes; no, he has 70 minutes and uses only 64 minutes.

August ....page 102

1. 5 times
2. August 23
3. Tuesday
4. August 7
5. 9 days

6. August 1
7. August 28
8. August 24
9. Friday
10. 3 days

Hummingbirds ....**page 103**

1. 11,400 beats
2. 136,800 beats
3. 2,700 feet
4. 4,800 beats
5. 6,909 flowers
6. 7,000 calories; 711 more needed
7. 7,380 calories in one hour
8. 154 species

Ready, Set, Go! ....**page 104**

Table 1

1. 6 times
2. 2 seconds faster
3. blink20 times and write name
4. 3 times
5. 45 seconds

Round About ....**page 105**

1. 161	**2.** 528	**3.** 251	**4.** 615
5. 361	**6.** 49	**7.** 287	**8.** 598
9. 515	**10.** 84	**11.** 666	**12.** 364
13. 307	**14.** 567	**15.** 96	**16.** 168

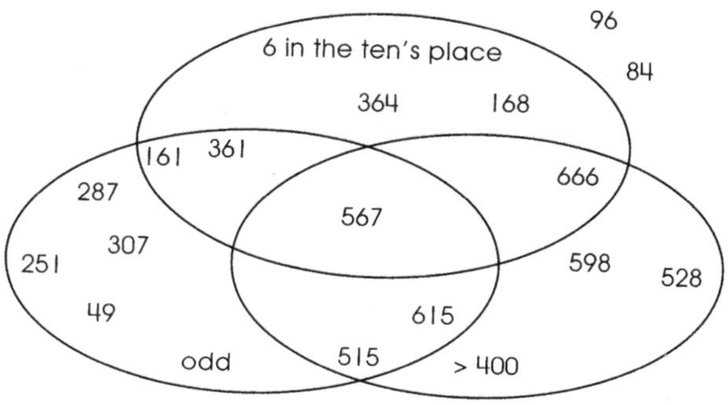

Line Up! ....**page 106**

1. The temperature increases as it reaches the middle of the day and decreases as the day wanes; This probably occurs because the light of the sun is more intense at midday, and therefore the temperature rises.

2.

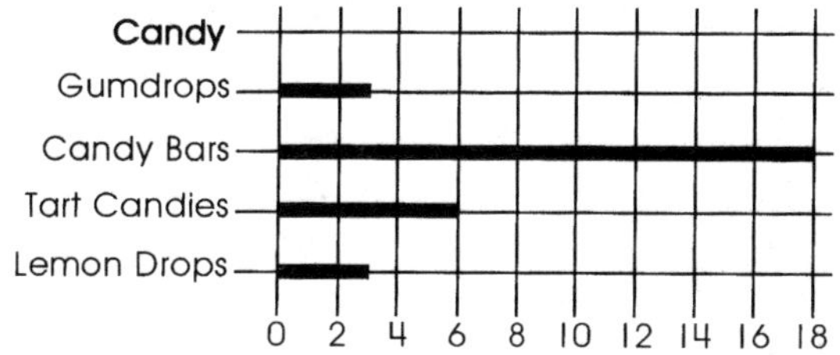

Softball ....**page 107**

1. Hedgehogs, Kangaroos, Cheetahs, Flamingos
2. Hedgehogs
3. Cheetahs; 25
4. Ferrets; 1
5. 24 points
6. Hedgehogs median = 6; mean = 6
7. Flamingos median = 9; mean = 10
8. **a.** 16; **b.** 8; **c.** 31; **d.** 1; **e.** 30; **f.** 18; **g.** 24; **h.** 6
9. no; The Hedgehogs won the tournament with a total of 18 points. Four teams had more points. Winners are determined by games won, not overall points.

Game Marathon ....**page 108**

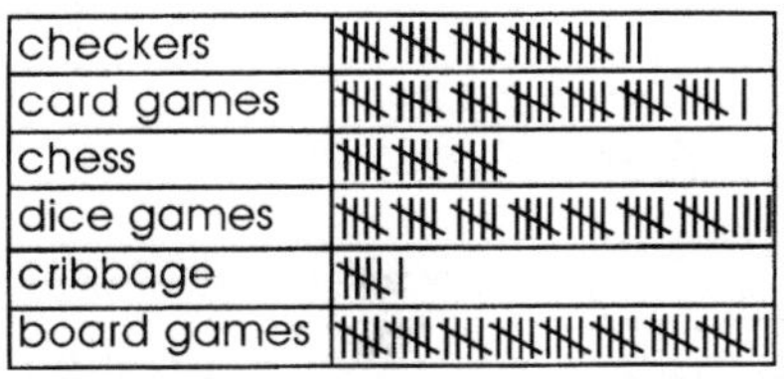

checkers	27
card games	36
chess	15
dice games	39
cribbage	6
board games	42

1. board games
2. 30 fewer games
3. card games
4. cribbage and chess
5. 12 more games

Ladybugs **109–110**

1. No; she found 10 with 2 spots. This and zero were the lowest.

2. 6 spots
3. 0 and 2 spots
4. yes; 4, 8, 12, and 14 spots
5. 20 more 10-spotted ladybugs

1. 27	2. 7	3. 14	4. 16
5. 56	6. 5	7. 64	8. 32
9. 12	10. 24	11. 9	12. 25
13. 2	14. 18	15. 8	16. 30
17. 20	18. 81	19. 42	20. 3
21. 6	22. 4	23. 10	24. 1
25. 15	26. 35	27. 362	8. 0

Ladybugs in the north have more spots to absorb more warmth from sunshine.

Probable Outcomes..........................page 111

1. 1 out of 8—⅛
2. 1 out of 8—⅛
3. 4 out of 8—½
4. 4 out of 8—½
5. 1 out of 8—⅛
6. 4 out of 8—4/8 or ½
7. blue, red, yellow, and green
8. red and yellow
9. green
10. 5 out of 8—⅝
11. 3 out of 8—⅜
12. 5 out of 8—⅝

Humpback Whales............................page 112

1. 35,200 krill
2. 34 feet
3. 10,000 miles
4. 1,000 whales
5. 360 feet
6. She would save 40¢ per month; she would save $4.60 in one year
7. 5 tons

Movie Night..page 113

1. He must leave by 5:15.
2. He will save $2.25.
3. The movie will end at 8:05.
4. He would save $1.31.
5. Yes, he could purchase a medium deal.

Theme Park ....page 114

1. 655 more cars
2. a. individual passes
   b. family pass
   c. family pass
3. 10 hours
4. 12.683 km
5. 2,640 ft.; 880 yd.
6. $2.69 change; two park pencils